We'Moon 2019
Gaia Rhythms for Womyn

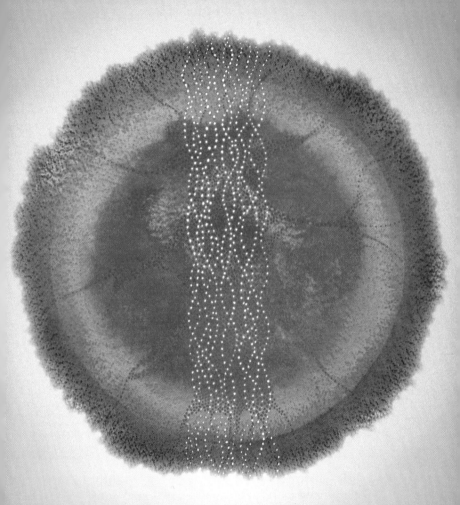

Spirit Nebula © Linda James 2014

Fanning the Flame

38th Edition of We'Moon

published by

Mother Tongue Ink

We'Moon 2019: Gaia Rhythms for Womyn
Spiral, Sturdy Paperback Binding, Unbound & Spanish Editions
© Mother Tongue Ink 2018

Mother Tongue Ink
Estacada, OR 97023
All Correspondence:
P.O. Box 187, Wolf Creek, OR 97497
www.wemoon.ws

We'Moon Founder/Crone Editor: Musawa, *Special Editor:* Bethroot Gwynn
We'Moonagers: Sue Burns, Barb Dickinson, *Graphic Design:* Sequoia Watterson
We'Moon Creatrix/Editorial Team: Bethroot Gwynn, Sequoia Watterson, Sue Burns, Leah Markman, Barb Dickinson, *Production Coordinator:* Barb Dickinson *Production Assistant & Retail Sales:* Leah Markman *Proofing*: EagleHawk, Sandra Pastorius, Kathryn Henderspn, Becky Bee, Amber Torrey—Iztåak Zòchitl *Promotion:* Leah Markman, Sue Burns, Susie Schmidt, Barb Dickinson *Accounts Manager:* Sue Burns *Order Fulfillment:* Susie Schmidt, *Shipping Assistant:* Dana Page

This eco-audit applies to all We'Moon 2019 *products:*

Hansol Paper **Environmental Benefits Statement:**
We'Moon 2019 is printed on Hansol paper using 60% recycled content: 50% pre-consumer waste, 50% post-consumer waste, with Solvent-free Soy and Vegetable Based inks with VOC levels below 1%.
By using recycled fibers instead of virgin fibers, we saved:
98 fully grown trees
39,250 gallons of water
27 million BTUs of energy
2,314 pounds of solid waste
6,925 pounds of greenhouse gasses

As a moon calendar, this book is reusable: every 19 years the moon completes a metonic cycle, returning to the same phase, sign and degree of the zodiac.

We'Moon is printed in South Korea by Sung In Printing America on recycled paper using low VOC soy-based inks.

Green America APPROVED FOR PEOPLE AND PLANET

Order directly from Mother Tongue Ink
To Order see p. 233. Email: weorder@wemoon.ws
Retail: 877-693-6666 or 541-956-6052 Wholesale: 503-288-3588

We'Moon 2019 Datebooks: • $21.95
Spiral ISBN: 978-1-942775-13-5
Sturdy Paperback ISBN: 978-1-942775-14-0
Unbound ISBN: 978-1-942775-15-7
Spanish Edition ISBN: 978-1-942775-16-4
In the Spirit of We'Moon • $26.95
Paperback ISBN: 978-1-890931-75-9
Preacher Woman for the Goddess • $16
Paperback ISBN: 978-1-890931-12-6

The Last Wild Witch • $9.95
Paperback ISBN: 978-1-890931-94-0
Other *We'Moon 2019* **Products:**
We'Moon on the Wall • $16.95
ISBN: 978-1-942775-17-1
Greeting Cards (6-Pack) • $11.95
ISBN: 978-1-942775-18-8
Organic Cotton Tote • $13
We'Moon Cover Poster • $10

2019

JANUARY
S	M	T	W	T	F	S
		1	2	3	4	**5**
6	7	8	9	10	11	12
13	14	15	16	17	18	19
20	21	22	23	24	25	26
27	28	29	30	31		

FEBRUARY
S	M	T	W	T	F	S
					1	2
3	**4**	5	6	7	8	9
10	11	12	13	14	15	16
17	18	**19**	20	21	22	23
24	25	26	27	28		

MARCH
S	M	T	W	T	F	S
					1	2
3	4	5	**6**	7	8	9
10	11	12	13	14	15	16
17	18	19	**20**	21	22	23
24	25	26	27	28	29	30
31						

APRIL
S	M	T	W	T	F	S
	1	2	3	4	**5**	6
7	8	9	10	11	12	13
14	15	16	17	18	**19**	20
21	22	23	24	25	26	27
28	29	30				

MAY
S	M	T	W	T	F	S
			1	2	3	**4**
5	6	7	8	9	10	11
12	13	14	15	16	17	**18**
19	20	21	22	23	24	25
26	27	28	29	30	31	

JUNE
S	M	T	W	T	F	S
						1
2	**3**	4	5	6	7	8
9	10	11	12	13	14	15
16	**17**	18	19	20	21	22
23	24	25	26	27	28	29
30						

JULY
S	M	T	W	T	F	S
	1	**2**	3	4	5	6
7	8	9	10	11	12	13
14	15	**16**	17	18	19	20
21	22	23	24	25	26	27
28	29	30	**31**			

AUGUST
S	M	T	W	T	F	S
				1	2	3
4	5	6	7	8	9	10
11	12	13	14	**15**	16	17
18	19	20	21	22	23	24
25	26	27	28	29	**30**	31

SEPTEMBER
S	M	T	W	T	F	S
1	2	3	4	5	6	7
8	9	10	11	12	**13**	14
15	16	17	18	19	20	21
22	23	24	25	26	27	**28**
29	30					

OCTOBER
S	M	T	W	T	F	S
		1	2	3	4	5
6	7	8	9	10	11	12
13	14	15	16	17	18	19
20	21	22	23	24	25	26
27	28	29	30	31		

NOVEMBER
S	M	T	W	T	F	S
					1	2
3	4	5	6	7	8	9
10	11	**12**	13	14	15	16
17	18	19	20	21	22	23
24	25	**26**	27	28	29	30

DECEMBER
S	M	T	W	T	F	S
1	2	3	4	5	6	7
8	9	10	**11**	12	13	14
15	16	17	18	19	20	21
22	23	24	**25**	26	27	28
29	30	31				

● = NEW MOON, PST/PDT   ◯ = FULL MOON, PST/PDT

Green Heart
© Bettina 'Star-Rose' Madini 2011

3

COVER NOTES

Brigid—The Goddess of Inspiration © Emily Balivet 2010

One of the most revered Goddesses of all time, Brigid is a mother, poetess, gardener, a resourceful healer, a fierce warrior, and protector of children. She is a fire Goddess and giver of Light. She draws from her cauldron and offers us the spark of creativity and the fiery will to take action. May she bless us all with divine intentionality!

Canyon Morning © Gael Nagle 2004

This piece evokes the deliciousness of the sun and the moment, of taking time to be aware and to relish. Gael is a batik artist. She paints melted beeswax onto cloth, submerges it in successive dye baths, then boils the wax out after each primary color is finished. Each piece takes months to complete. Her work is inspired by an appreciation of the human experience in the natural world.

DEDICATION

Every year, we donate a portion of our proceeds to an organization doing good work that resonates with our theme—bringing positive change to the world and to the lives of women. Fanning the Flame, this year's We'Moon theme, calls for direct action, fiery fury, and transformative interruption. We dedicate this edition of We'Moon to The Guerrilla Girls.

The Guerrilla Girls, founded in 1985, are anonymous feminist activist artists. They wear gorilla masks in public and use facts, humor and outrageous visuals to expose gender and ethnic bias as well as corruption in politics, art, film, and pop culture. GG believe in intersectional feminism that fights discrimination and supports human rights for all people and all genders. They have done hundreds of projects (posters, actions, books, videos, stickers) all over the world, including Bilbao, Iceland, Istanbul, London, Los Angeles, Mexico City, New York, Rotterdam, Sao Paolo, and Shanghai. They also do

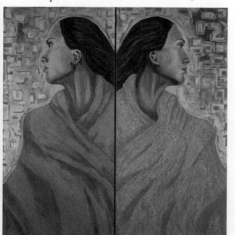

interventions and exhibitions at museums, blasting them on their own walls for their bad behavior and discriminatory practices. Guerilla Girls could be anyone. They are everywhere. They encourage: *More creative complaining!! More interventions!! More resistance!!*

Email gg@guerrillagirls.com, learn more at guerrillagirls.com

Sue Burns © Mother Tongue Ink 2018

Espejo / Mirror © *Carmen R. Sonnes 2006*

TABLE OF CONTENTS
INTRODUCTION

MOON CALENDAR: FANNING THE FLAME

APPENDIX

WE'MOON 2019 FEATURE WRITERS:

We'Moon Wisdom: Musawa; **Astrologers:** Rhea Wolf, Heather Roan Robbins, Sandra Pastorius, Gretchen Lawlor, Susan Levitt, Mooncat!, Beate Metz; **Introduction to the Theme:** Bethroot Gwynn; **Holy Days:** Susa Silvermarie; **Lunar Phase Card:** Susan Baylies; **Herbs:** Sue Burns; **Tarot:** Leah Markman

What Is *We'Moon*? A Handbook in Natural Cycles

We'Moon: Gaia Rhythms for Womyn is more than an appointment book, it's a way of life! We'Moon is a lunar calendar, a handbook in natural rhythms, and a collaboration of international womyn's cultures. Art and writing by wemoon from many lands give a glimpse of the great diversity and uniqueness of a world we create in our own images. We'Moon is about womyn's spirituality (spirit-reality). We share how we live our truths, what inspires us, and our connection with the whole Earth and all our relations.

Wemoon **means "we of the moon."** The Moon, whose cycles run in our blood, is the original womyn's calendar. We use the word "wemoon" to define ourselves by our primary relation to the cosmic flow, instead of defining ourselves in relation to men (as in wo*man* or fe*male*). We'Moon is sacred space in which to explore and celebrate the diversity of she-ness on Earth. We'Moon is created by, for and about womyn: in our image.

We'Moon culture exists in the diversity and oneness of our experiences as wemoon. We honor both. We come from many different ways of life. At the same time, as wemoon, we share a common mother root. As makers of We'Moon, we are delighted when wemoon from varied backgrounds contribute art and writing. We'Moon does not support or condone cultural appropriation (taking what belongs to others) or cultural fascism (controlling artistic expression). We do not knowingly publish oppressive content of any kind. We invite you to share your work with respect for both cultural integrity and creative inspiration.

We'Moon is dedicated to amplifying the images and voices of wemoon from many perspectives and cultures. We are fully aware that we live in a racist patriarchal society. Its influences have permeated every aspect of society, including the very liberation movements committed to ending oppression. Feminism is no exception—historically and presently dominated by white women's priorities and experiences. We seek to counter these influences in our work. Most of us in our staff group are lesbian or queer—we live outside the norm. At the same time, we are mostly womyn who benefit from white privilege. We seek to make We'Moon a safe and welcoming place for all wimmin, especially for women of color (WOC) and others marginalized by the mainstream. We are eager to publish more words and images depicting people of color created by WOC. We encourage more WOC to submit their creative work to We'Moon for greater inclusion and visibility (see p. 236).

Lunar Rhythms: Everything that flows moves in rhythm with the Moon. She rules the water element on Earth. She pulls on the ocean's tides, the weather, female reproductive cycles, and the life fluids in plants, animals and people. She influences the underground currents in Earth energy, the mood swings of mind, body, behavior and emotion. The Moon's phases reflect her dance with Sun and Earth, her closest relatives in the sky. Together, these three heavenly bodies weave the web of light and dark into our lives.

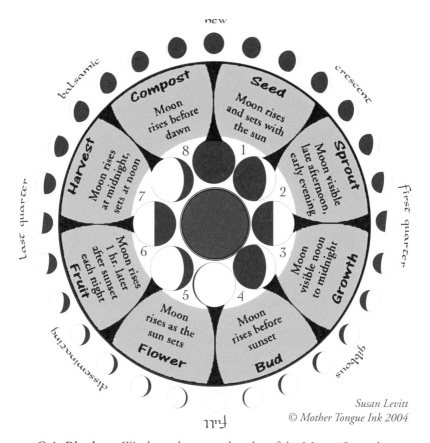

new

balsamic

crescent

Compost
Moon rises before dawn

Seed
Moon rises and sets with the sun

Harvest
Moon rises at midnight, sets at noon

Sprout
Moon visible late afternoon, early evening

last quarter

first quarter

Fruit
Moon rises 1 hr. later after sunset each night

Growth
Moon visible noon to midnight

Flower
Moon rises as the sun sets

Bud
Moon rises before sunset

disseminating

gibbous

full

Susan Levitt
© *Mother Tongue Ink 2004*

Gaia Rhythms: We show the natural cycles of the Moon, Sun, planets and stars as they relate to Earth. By recording our own activities side by side with those of other heavenly bodies, we may notice what connection, if any, there is for us. The Earth revolves around her axis in one day; the Moon orbits around the Earth in one month (29½ days); the Earth orbits around the Sun in one year. We experience each of these cycles in the alternating rhythms of day and night, waxing and waning, summer and winter. The Earth/Moon/Sun are our inner circle of kin in the universe. We know where we are in relation to them at all times by the dance of light and shadow as they circle around one another.

The Eyes of Heaven: As seen from Earth, the Moon and the Sun are equal in size: "the left and right eye of heaven," according to Hindu (Eastern) astrology. Unlike the solar-dominated calendars of Christian (Western) patriarchy, We'Moon looks at our experience through both eyes at once. The **lunar eye of heaven** is seen each day in the phases of the Moon, as she is both reflector and shadow, traveling her 29½-day path around the Earth in a "Moon" Month (from each new moon to the next, 13 times in a lunar year). Because Earth is orbiting the Sun at the same time, it takes the Moon 27$^1/_3$ days to go through all the signs of the Zodiac—a sidereal month. The **solar eye of heaven** is apparent at the turning points in the Sun's cycle. The year begins with Winter Solstice (in the Northern Hemisphere), the dark renewal time, and journeys through the full cycle of seasons and balance points (solstices, equinoxes and the cross-quarter days in between). The **third eye** of heaven may be seen in the stars. Astrology measures the cycles by relating the Sun, Moon and all other planets in our universe through the backdrop of star signs (the zodiac), helping us to tell time in the larger cycles of the universe.

Measuring Time and Space: Imagine a clock with many hands. The Earth is the center from which we view our universe. The Sun, Moon and planets are like the hands of the clock. Each one has its own rate of movement through the cycle. The ecliptic, a 17° band of sky around the Earth within which all planets have their orbits, is the outer band of the clock where the numbers are. Stars along the ecliptic are grouped into constellations forming the signs of the zodiac—the twelve star signs are like the twelve numbers of the clock. They mark the movements of the planets through the 360° circle of the sky, the clock of time and space.

Whole Earth Perspective: It is important to note that all natural cycles have a mirror image from a whole Earth perspective—seasons occur at opposite times in the Northern and Southern Hemispheres, and day and night are at opposite times on opposite sides of the Earth as well. Even the Moon plays this game—a waxing crescent moon in Australia faces right (☾), while in North America, it faces left (☽). We'Moon uses a Northern Hemisphere perspective regarding times, holy days, seasons and lunar phases. Wemoon who live in the Southern Hemisphere may want to transpose descriptions of the holy days to match seasons in their area. We honor a whole Earth cultural perspective by including four rotating languages for the days of the week, from different parts of the globe.

Whole Sky Perspective: It is also important to note that all over the Earth, in varied cultures and times, the dome of the sky has been

interacted with in countless ways. The zodiac we speak of is just one of many ways that hu-moons have pictured and related to the stars. In this calendar we use the Tropical zodiac, which keeps constant the Vernal Equinox point at 0° Aries. Western astrology primarily uses this system. Vedic or Eastern astrology uses the Sidereal zodiac, which bases the positions of signs relative to fixed stars, and over time the Vernal Equinox point has moved about 24° behind 0° Aries.

Soul Card # 21 © Deborah Koff-Chapin 1995

Musawa
© *Mother Tongue Ink 2008*

HOW TO USE THIS BOOK
Useful Information about We'Moon

Time Zones: All aspects are in Pacific Standard/Daylight Time, with the adjustment for GMT and EDT given at the bottom of each page. To calculate for other areas, see "World Time Zones" (p. 234).

Signs and Symbols at a Glance is an easily accessible handy guide that gives brief definitions of commonly used astrological symbols (p. 229).

Pages are numbered throughout the calendar to facilitate cross referencing. See Table of Contents (p. 5) and Contributor Bylines and Index (pp. 192–203). The names of the days of the week and months are in English with four additional language translations: Laadan, Farsi, Spanish and Mandarin.

Lunar Calendar Moon Theme Pages mark the beginning of each moon cycle with a two-page spread near the new moon. Each *Moon Page* is numbered with Roman numerals followed by the theme for that month (e.g., **IV: Love Sparks**) and contains the dates of that *Moon's* new and full moon and solar ingress.

Year at a Glance Calendars are on pp. 3 and 230. **Month at a Glance Calendars** can be found on pp. 214–225 and include daily lunar phases. Susan Baylies' **Lunar Phase Card** features the moon phases for the entire year on pp. 226–227.

Annual Astro Portraits: To find your astrological portrait for the year by Rhea Wolf, turn to "Intro to the Astro Glances" (p. 19).

Holy Days: There is a two-page Holy Day spread for all equinoxes, solstices and cross-quarter days. These include feature writings by Susa Silvermarie, accompanied by additional art and writing.

Planetary Ephemeris: Exact planetary positions for every day are given on pp. 208–213. These ephemerides show where each planet is in a zodiac sign at noon GMT, measured by degree in longitude in Universal Time.

Asteroid Ephemeris: Exact positions of asteroids for every ten days are given for sixteen asteroids in the zodiac at midnight GMT on p. 207.

Astrology Basics (Refer to sample calendar page, p. 10)

Planets: Planets are like chakras in our solar system, allowing for different frequencies or types of energies to be expressed. See Mooncat's article (pp. 204–205) for planetary attributes.

Signs: The twelve signs of the zodiac are a mandala in the sky, marking off 30° segments in the 360° circle around the earth. Signs show major shifts in planetary energy through the cycles. (See **Constellations of the Zodiac** p. 228)

Glyphs: Glyphs are the symbols used to represent planets and signs. See "Signs and Symbols at a Glance" (p. 229).

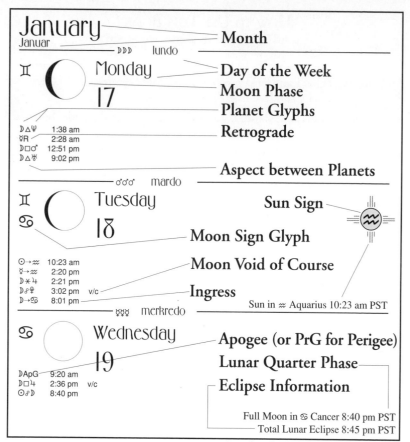

January
Januar

ᗪᗪᗪ   lundo

♊   Monday — **Day of the Week**
      Moon Phase
      Planet Glyphs
17   **Retrograde**

☽△♆   1:38 am
☿R   2:28 am
☽□♂   12:51 pm
☽△♅   9:02 pm

    Aspect between Planets

♂♂♂   mardo

♊   Tuesday   **Sun Sign**
♋   18   **Moon Sign Glyph**

☉→♒   10:23 am   **Moon Void of Course**
☿→♒   2:20 pm
☽⚹♃   2:21 pm
☽♂♀   3:02 pm v/c   **Ingress**
☽→♋   8:01 pm
            Sun in ♒ Aquarius 10:23 am PST

☿☿☿   merkredo

♋   Wednesday   **Apogee (or PrG for Perigee)**
      19   **Lunar Quarter Phase**
            Eclipse Information

☽ApG   9:20 am
☽□♃   2:36 pm v/c
☉♂☽   8:40 pm

            Full Moon in ♋ Cancer 8:40 pm PST
            Total Lunar Eclipse 8:45 pm PST

Month

Sample calendar page for reference only

Sun Sign: The Sun enters a new sign once a month (on or around the 21st), completing the whole cycle of the zodiac in one year. The sun sign reflects qualities of your outward shining self.

Moon Sign: The Moon changes signs approximately every 2 to 2 1/2 days, going through all twelve signs of the zodiac every 27 1/3 days (the sidereal month). The Moon sign reflects qualities of your core inner self.

Moon Phase: Each calendar day is marked with a graphic representing the phase of the Moon.

Lunar Quarter Phase: At the four quarter-points of the lunar cycle (new, waxing half, full and waning half moons), we indicate the phase, sign and exact time for each. These points mark off the "lunar week."

Day of the Week: Each day is associated with a planet whose symbol appears in the line above it (e.g., ᗪᗪᗪ is for Moon: Moonday, Monday, Luna Day, lundi, lunes). The names of the days of the week are displayed prominently in English with translations appearing in the line above them.

10

The languages—Laadan, Farsi, Spanish and Mandarin—rotate weekly, in this order, throughout the calendar.

Eclipse: The time of greatest eclipse is given, which is near to, but not at the exact time of the conjunction ($\odot \, \sigma \, \mathcal{D}$) or opposition ($\odot \, \mathcal{S} \, \mathcal{D}$). Locations from where eclipses are visible are also given. See "Eclipses" (p. 22).

Aspects ($\square \, \triangle \, \mathcal{S} \, \sigma \, \maltese \, \pi$): These show the angle of relation between different planets. Daily aspects provide something like an astrological weather forecast, indicating which energies are working together easily and which combinations are more challenging. See "Signs and Symbols at a Glance" (p. 229).

Ingresses ($\rightarrow$): When the Sun, Moon and planets move into new signs.

Moon "Void of Course" ($\mathcal{D}$ **v/c**): The Moon is said to be "void of course" from the last significant lunar aspect in each sign until the Moon enters a new sign. This is a good time to ground and center yourself.

Super Moon: A Super Moon is a New or Full Moon that occurs when the Moon is at or within 90% of perigee, its closest approach to Earth. On average, there are four to six Super Moons each year. Full Super Moons could appear visually closer and brighter, and promote stronger tides. Personally, we may use the greater proximity of Super Moons to illuminate our inner horizons and deepen our self-reflections and meditations.

Apogee (ApG): The point in the Moon's orbit that is **farthest** from Earth. At this time, the effects of transits may be less noticeable immediately, but may appear later. Also, **Black Moon Lilith**, a hypothetical second foci or center point of the elliptical orbit the Moon makes around the Earth, will be conjunct the Moon.

Perigee (PrG): The point in the Moon's orbit that is **nearest** to Earth. Transits with the Moon, when at perigee, will be more intense.

Lunar Nodes: The most Northern and Southern points in the Moon's monthly cycle when it crosses the Sun's ecliptic or annual path, offering to integrate the past (South) and future (North) directions in life.

Aphelion (ApH): The point in a planet's orbit that is **farthest** from the Sun. At this time, the effects of transits (when planets pass across the path of another planet) may be less noticeable immediately, but may appear later.

Perihelion (PrH): The point in a planet's orbit that is **nearest** to the Sun. Transits with planets, when they are at perihelion, will be more intense.

Direct or Retrograde (D or R): These are times when a planet moves forward (D) or backward (R) through the signs of the zodiac (an optical illusion, as when a moving train passes a slower train that appears to be going backward). When a planet is in direct motion, planetary energies are more straightforward; in retrograde, planetary energies turn back in on themselves and are more involuted. See "Mercury Retrograde" (p. 22).

Electric Universe ▫ *Susan Solari 2016*

ASTROLOGICAL OVERVIEW: 2019

Crank up the heat. We need to be on fire this year, but whether we feel the hearth fire of the gathering collective, the fire of a combustion engine, or just burn the house down—that will be up to us and how we use the aspects.

So, let's bring that solar spark into our hearts and catch fire in the best of ways. We'll need a spark as the social planets Jupiter and Saturn, and most of the outer planets that hold the cauldron of our present reality, head into grounded, concrete, territorial, earth signs. Like a mountain, earth signs bring energy that is hard to get moving, but once the earthquake begins the momentum can be unstoppable. Saturn and Pluto are both in determined Capricorn as 2019 begins, Uranus enters grounded Taurus on **March 6**; Jupiter enters Capricorn in **December**. But we can take this pragmatic strength within us, add our passion, and really make a difference.

Of the outer planets, Neptune is the only one that will not be in an earth sign, but in the intuitive, if confusing, water sign Pisces. While this encourages our spiritual lives to flow and allows our intuition and magic to be a real resource for us, it may also keep religion a point of illusion and contention in world politics.

Jupiter, Saturn and Pluto will conjunct in 2020, right before the USA presidential election. When two or more planets come together, like the Sun and Moon at the New Moon, they begin a new cycle.

This three-way conjunction in 2020 will offer us a chance to reset our social and political world. What we do in 2019 can wrap up the old cycle, will directly impact the direction of that new cycle and influence the world's politics in the years ahead.

Passions run high as 2019 begins. Mars inhabits its own sign of Aries, and we feel the need to fight the good fight. Venus begins the year in Mars-ruled Scorpio, and at its greatest Western elongation, bright in the morning sky and strong in our hearts. A partial solar eclipse on **January 5,** balanced right between Saturn and Pluto, can help us define our work for the next year, followed by a lunar eclipse on **January 20** which asks us to balance our personal life with our work in the community.

But all this passion takes clarity, and our perception may be clouded by some illusions as expansive Jupiter squares Neptune on **January 13, June 16**, and **September 21**. We have to step away from either a desire to escape the world, from the illusion that there is no hope, or the illusion that everything is just fine; clear the veils of illusion and see what truly is. This will be as true of our romantic relationships as it is of our politics.

To help us see more clearly, a subtle Saturn-Neptune sextile makes three passes, on **January 31, June 18**, and **November 8**, and can bring discipline to our spiritual life and our creative process. Our own disciplines (a positive use of Saturn in Capricorn), those we choose rather than those imposed from the outside, can stabilize us and help us change the collective narrative.

Chiron dipped briefly into Aries in 2018 then retrograded back to Pisces, but on **February 18**, Chiron enters Aries and stays there through 2027. Chiron helps us see our wounds, heal them, and use that new understanding to help others. In Aries, Chiron asks us to create a healthy revolution, to come alive and fight for good causes without wounding others. Because Chiron exposes our wounds, we may see a resurgence of machismo and need to claim our passionate Amazon, and search for a way to move forward that doesn't polarize but is direct, clear, and to the point.

Change-master Uranus enters earthy, stubborn, matter-focused Taurus on **March 6** for the next eight years. Uranus was last in

Taurus at the end of the Great Depression and the beginning of the New Deal; here it asks us to transform our relationship to our earthly resources, our money, our bodies.

By the end of **April**, Jupiter, Pluto, and Saturn all turn retrograde, and Neptune retrogrades on the Summer Solstice. So **May, June,** and **July** bring us back to review work we thought was already accomplished. We need to be patient and exercise our new muscles and new collaborations to keep the momentum moving forward. In early **June**, Jupiter squares Neptune again and could bring discouragement that we will need to break, burn off that fog in the sunlight so we can see what's really going on.

A solar eclipse on **July 2**, followed by a lunar eclipse on **July 16,** insists that we really look at what nurtures us, versus what we have to pare away. These eclipses may hammer on gender role issues, so we each need to hold a balance between Mars and Venus energies within, between Moon and Saturn, mother and father—both in personal relationships and in political realms.

Jupiter turns direct in **August**, Saturn in **mid-September**, and Pluto on **October 2**, and we feel the work pick up again. We can approach new terrain in our work, open to new programmatic and organizational possibilities.

On **November 11**, retrograde Mercury in Scorpio transits across the face of the sun, as the moon does during an eclipse. Although Mercury circumnavigates the sun 3 times a year, this direct alignment only happens 13 times per century and can make conscious what's been hidden within. Pay attention to fresh insights and new information.

Jupiter enters Capricorn on **December 2** and trines Uranus on **December 15**, nudging us to break free from old patterns and take concrete steps forward. We may need to change our relationship to our possessions and finances to do so in a stable and healthy way.

2019 ends with a solar eclipse/New Moon conjunct, liberating Jupiter at 4° of Capricorn, which asks us to live out the best of Capricornian mythology, to bring that sunlight into our souls to energize our search for a dream, to find people who share our dreams, and walk them to the tops of the mountains in the year ahead.

Heather Roan-Robbins © Mother Tongue Ink 2018

Sun Cycles and Aspects

Sunna, the roaming Nordic Goddess, lit up our world as she restlessly rode her chariot searching for home. Aine, Irish Goddess of summer, abundance and sovereignty, galloped as a red horse across the sky. Bast, Egyptian lion Goddess of the quickening solar rays, protected her people. Ameratsu's rising Sun still illuminates the Japanese flag. Around the world the Sun, reflected through the divine feminine, brings

Gaia Sagrada □ *Heather Brunetti 2017*

life, light, abundance, vitality, and infuses personal sovereignty. She lights up our life.

The Sun organizes our solar system and marks our days. All planets in our charts are illuminated only through their relationship to the Sun. Sun's energy fluctuates; it burns at different levels of intensity and turns the volume up or down on events here on earth. Solar activity cycles peak on the average about every 11 years, but the cycle is unpredictable, and can be 9 to 14 years apart and of greater or lesser intensity. We can't draw up tables ahead of time like all other planetary patterns, but we can find out what the Sun is doing today, and was doing the day we were born. (You can look up that information at www.spaceweather.com)

It's helpful to know what the Sun is doing because her cycles synchronize with fluctuations here on earth; her pattern affects tree growth, crops, health cycles, and adjusts the volume on earthly events. Our politics heat up when the Sun flares and storms. We're hungry for new answers. Earthlings become more politically volatile so a spark can light a bonfire. Solar activity peaked around the American, French, and Russian revolutions, the late 1960's, and in 2001. Social revolutions tend to switch directions with each new cycle, oscillating between conservative and progressive peaks.

Obama came in as the Sun grew active, promising a wave of change which never quite materialized, as the Sun went deeply quiet right after his election. The recent conservative 2013 sunspot maximum was, with occasional spikes, the smallest in a hundred years.

The Sun is predicted to quiet down through 2019 then build again towards a progressive solar maximum around 2022–2025. In the years when the Sun is quiet, it's hard to light a match and spark major political or cultural change. The culture tends to work with the answers in place, whether we like it or not. But we can network, produce art that brings questions, educate and organize. It's time to push our spiritual lives. Develop philosophy and weave connections, lobby for legal changes, change hearts, and get good politicians in office. Educate and have a life. Get the building blocks in place to activate when the Sun flares.

Solar aspects: The Sun progresses about a degree a day and makes its way through the 360° of the zodiac every year effecting different aspects. Solar aspects involve special and energetic relationships between the Sun and planets. We feel solar aspects building for 9 days before they peak, and resonating up to 7 days afterwards. The qualities of whatever planet the Sun aspects tinge the nature of this interaction, and all that pours through the filter of the Sun's sign.

Challenging aspects—square, semi-square, opposition, quincunx, semi sex-tile— can create complications, but often give us the energy to tackle a problem. **Supportive aspects**—conjunction, trine, sextile—assist, but are easier to ignore.

Moon phases are really a Sun-Moon relationship. The Sun energizes all Moon things: the public, nurturance, home, intimacy, habit, and the tides of emotion. Supportive aspects ease the emotional realm; challenging aspects can create tension between how we feel and the events around us.

Sun and Mercury are never more than a sign apart, so their aspects are limited, yet they energize our mind-body connection. When there's tension between Mercury and the Sun, willfulness and reason tug at one another. When the Sun and Mercury constellate, our minds buzz, and thought easily becomes action.

Sun and Venus are never more than two signs apart; when the Sun and Venus work together they energize love, affection, sensuality aesthetics, compassion, cooperation. When these two are at odds, our emotions can tug at our life purpose; it roils our heart.

Mars is the first planet past earth and can form all aspects with the Sun. When the Sun challenges Mars, we become easily inflamed—tempers spark, mechanical problems and machismo get obstreperous. Avoid dueling willpowers. When Mars and the Sun constellate, it energizes our physical vitality, heroism, sexuality, swagger, and endurance.

Jupiter, the largest planet, expands whatever it touches. When the Sun challenges Jupiter, we tend to overdo, eat too much, say too much, have trouble holding secrets, and can over-enable with our generosity. When the Sun and Jupiter constellate, generosity is more abundant and benevolent.

Sun-Saturn builds and restricts; it offers skeletal structure. When the Sun and Saturn clash, our responsibilities lean in—our pressures, schedules feel tight; attitudes turn grim, age aches, rules and regulations itch—but we can hold our ground. When the Sun and Saturn constellate, it energizes our maturity, firm boundaries, and organizational skills.

Sun-Uranus spins backwards and sideways; one pole always points towards the Sun, creating an unusual electrical field. When the Sun challenges Uranus, it energizes our restlessness, anxiety, our contrary personality. Electrical and mechanical devices act up. When the Sun and Uranus constellate, it energizes our ingenuity, ability to change, adapt, experiment.

Ephemeral **Neptune** speaks of non-ordinary reality, spirituality, creativity, escapism, addictions. When the Sun irritates Neptune, our daydreams clash with our work, and we may want to nap, escape, or step out of ordinary reality. When they constellate, our dreams walk into our days, and our intuition and imagination are energized.

Pluto symbolizes power and transformation. When the Sun challenges Pluto, our life force responds to tough memories and grim realities. When the Sun and Pluto constellate, they energize our empowerment and remind us of our deepest priorities.

by Heather Roan Robbins
© Mother Tongue Ink 2018

Jaguar Guide *© Francene Hart 2008*

SKY MAP 2019

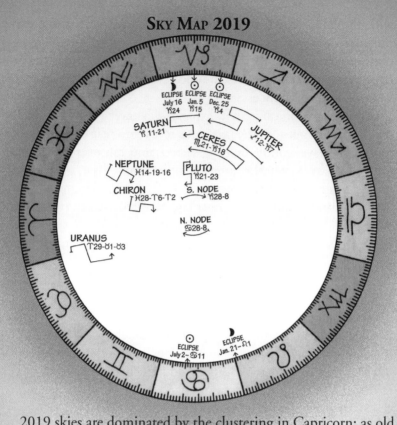

ECLIPSE
July 16
♑24

ECLIPSE
Jan. 5
♑15

ECLIPSE
Dec. 25
♑4

SATURN
♑ 11-21

CERES
♏21-♑18

JUPITER
♐12-♑7

NEPTUNE
♓14-19-16

PLUTO
♑21-23

CHIRON
♓28-♈6-♈2

S. NODE
♑28-8

N. NODE
♋28-8

URANUS
♈29-♉1-♉3

ECLIPSE
July 2- ♋11

ECLIPSE
Jan. 21- ♌1

2019 skies are dominated by the clustering in Capricorn; as old systems implode and fragment, we wisely and resolutely build local and personal autonomy. This steadies our resilience, and we become as Phoenix, ready to rise out of the ashes. As the 5 eclipses of 2019 jar Capricorn structures, we continue to wake out of any remaining complacency that all will return to as it once was.

Jupiter, Intrepid Explorer, peaks in influence in 2019—in Sagittarius until Dec 3; if we seek, we can discover reasons to hope for a better future. Jupiter aligning with Uranus in the last moments of Aries ignites a fresh wave of courageous innovation and radicalizing genius. Uranus then moves into Taurus for 7 years; we fare best if unencumbered by excessive attachment to our stuff. Experiment with local economies.

Neptune aligning with Capricorn continues to seed rich dreams, unencumbered by old realities. The Moon's nodes: all striving this year must be cyclic, balanced and sustained through frequent, soulful nourishing inbreath.

Gretchen Lawlor © Mother Tongue Ink 2018

Astrological Year at a Glance Introduction

In 2019, the revolution ignites within our hearts—the fiery center of love. The revolutionary heart reminds us that the truth of our power lives in connection and collaboration. And it is available in every moment; whenever we choose to give generously of our skills and to cooperate with others, we are birthing a new world.

This revolution is sometimes called *The Great Turning*, a phrase used by visionary teacher Joanna Macy to describe the shift we are making to a life-sustaining culture, and it has three main components: Holding Actions, Innovative Structures and Shifts of Consciousness.

As Saturn continues its 3-year cycle through Capricorn, coming closer to Pluto, we can slow destruction and diminish harm, even as things intensify, by engaging in "Holding Actions." Holding actions occur anytime we take a stand against the forces of greed and corruption. This has been an ongoing lesson during the long transit of Pluto in Capricorn, from 2008-2024. Saturn emboldens holding actions that block state violence or oil pipelines, as well as practical tasks, such as caring for those left behind by capitalism.

The second facet of *The Great Turning* arises as we build "Innovative Structures." Anytime we generously collaborate in power-with relationships, we are defying power-over systems of domination. New societal structures—or ancient ones coming back into prominence—are catalyzed by Uranus' 8-year stint in Taurus that began mid-2018. Taurus is the sign of stewardship and sustainability, while Uranus brings innovation. Open to the wild and listen—Gaia's voice encourages us to be inventive as we create communities of true abundance and plentiful resources.

Jupiter's dance through its home sign of Sagittarius for most of 2019 amplifies the third aspect of *The Great Turning*: "Shift of Consciousness." Sagittarius is the sign of the truth-seeker, and Jupiter's influence there expands our ability to see beyond the mundane and into the interconnected core of our creative, vibrant universe. As Marianne Williamson says, "...as our perception of an object changes, the object itself literally changes," meaning that as you shift how you see yourself and the world, you change yourself and the world. That's the revolutionary heart: all of us, waking up.

Rhea Wolf © Mother Tongue Ink 2018

To learn more about astrological influences for your sign, find your Sun and your Rising signs in the pages noted to the right.

Sun Cycling Through the Houses

The Astrological Sun is a primary symbol for the awareness of our unique human Self, and the creative expressions and passionate outpourings that we gift to the world. As Earth beings we circle around the Sun, year after year.

We may follow the transiting Sun's passage through the "Houses" (the twelve pie shaped sections) of our personal birth chart to attend to all areas of our life. Each house represents a basic facet of human experience from birth to death. The Sun spends about a month highlighting the affairs of each house, completing a full cycle in one year. Use the guide below to cultivate and enrich your Solar aspirations.

The **First House** represents our will to Be, and the experience of becoming an individual Self. From the moment of our first breath, we strengthen in chi, and gradually keep discovering, expressing and becoming more conscious of our greater Selves. (Fire)

In the **Second House** we delight in secure grounding with the Earth. We sustain our bodies through the use and appreciation of food, shelter, and material goods for living and thriving. (Earth)

The **Third House** highlights practical thought and our capacity to explore, learn and share communications. We are curious to understand and delight in the associations we make. Strengthen connections with peers, and team up. (Air)

The **Fourth House** concerns developing centered rootedness. We create home as a personal base of operation, as well as the psycho-spiritual home we carry within. Nurture deeper emotional expression with attention to family of origin. (Water)

In the **Fifth House** we let the adventures of life carry our aspirations and creative expression into fresh areas, and delight in embellishing our individuality. With heartfelt generosity we entertain children of our body and mind, including our inner child. (Fire)

Our **Sixth House** offers up self-improvement through practical work, service to others and disciplined training. Attend to the health and maintenance of your self and your community. Learn skillful techniques to sort the wheat of life from the chaff. (Earth)

The **Seventh House** is opposite from the First, and balancing relationships with others is highlighted. We have honed our best selves through the first six houses. Now we are ready to join in partnership, and polish the art of collaboration. (Air)

The **Eighth House** is concerned with joint resources, including our financial affairs and meeting needs for intimate connection and sexuality. As we merge deeply with others we may care passionately, hold tightly, then allow for dissolution to take place. (Water)

Ninth House priorities align our values, beliefs and our philosophy of life. As we further our education, we can tap into higher levels of inspiration, unfold in wisdom and develop our social conscience. (Fire)

The **Tenth House** summons us to the peak of our intentions and our chosen life work. Our task here is to take responsibility for building our reputation, and sharing our gifts and skills with meaningful livelihood. (Earth)

In the **Eleventh House** territory we fine tune our intellect with social skills. Interrelationships with groups and friendships open up the big picture. Choose seed visions for the next Sun cycle and set intentions with imagination and higher purpose. (Air)

In the **Twelfth House** lies our subconscious, where our dreams and affairs of the past reside and seek release. Alone, yet intuitively merged with collective identity, we reap our rewards, chase our shadows and accompany our muse. (Water)

Sandra Pastorius © Mother Tongue Ink 2018

To learn the dates that the Sun enters each of your Houses, refer to your Birth Chart for the Sign and degree on each house cusp, and then use an Ephemeris (p. 208) to find the corresponding date that the Sun enters each of the Houses. You may calculate your chart using: Astro.com, or contact Sandra for assistance: sandrapastorius@gmail.com

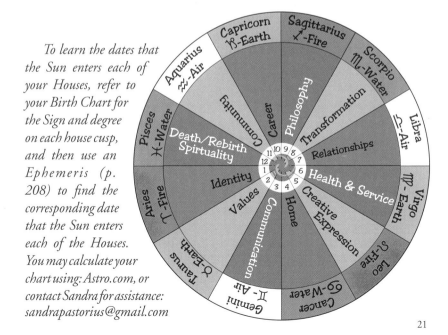

ECLIPSES: 2019

Solar and Lunar Eclipses occur when the Earth, Sun and Moon align at the Moon's nodal axis, usually four times a year, during New and Full Moons, respectively. The South (past) and North (future) Nodes symbolize our evolutionary path. Eclipses catalyze destiny's calling. Use eclipse degrees in your birth chart to identify potential release points.

January 5: Partial Solar Eclipse at 15° Capricorn reveals any need for recognition. Offer yourself acknowledgement and respect, let others appreciate your shine. Meet commitments consciously.

January 20: Total Lunar Eclipse at 0° Leo highlights our need to be seen and appreciated. When the light returns let your desires be known and share your best self. Enjoy the limelight.

July 2: Total Solar Eclipse at 10° Cancer stirs up our need for love. Urges for giving and receiving arise. Allow for receptivity, and offer affection with your heart showing.

July 16: Partial Lunar Eclipse at 24° Capricorn illuminates our need to matter. As the light returns, allow self-worth to radiate your true value. Ground positive input, and release the negative.

December 25: Annular Solar Eclipse at 4° Capricorn focuses on your needs for security. If vulnerabilities surface, hold to your center. As the light returns, choose courage and invite support.

MERCURY RETROGRADE: 2019

Mercury, planetary muse and mentor of our mental and communicative lives, appears to reverse its course three or four times a year. We may experience less stress during these periods by taking the time to pause and go back over familiar territory and give second thoughts to dropped projects or miscommunications. Breakdowns can help us attend to the safety of mechanics and mobility. It's time to "recall the now" of the past and deal with underlying issues. Leave matters that lock in future commitments until Mercury goes direct.

Mercury has three retrograde periods this year in water signs:

March 5–28: When Mercury retrogrades in Pisces, allow for reflective time into your past, and flesh out old grievances. Upgrade current concerns. Refresh your connections.

July 7–31: As Mercury retraces through Leo/Cancer, look for underlying meanings. Use creativity and verve to work out family discontents. Let your circle of support grow.

October 31–November 20: While Mercury is retrograde in Scorpio, take alone time for introspection and examine interpersonal commitments. Allow any fretting to shake up outworn attachments.

THE YEAR OF THE PIG 2019

The Year of the Pig begins on the second new Moon after Winter Solstice: February 4th. Pig year is a time of celebration, peace making, and enjoying the fine things in life, especially good food. Now is the year to take a vacation, relax and not work too hard. Kind actions are rewarded, and people can work together in harmony.

Poland: Pigs © Betty LaDuke 2005

Of the five Taoist elements—Fire, Earth, Metal, Water, and Wood—2019 is the year of the Earth Pig, the fortunate gold Pig. Under the influence of the golden Pig, we can find peace, happiness, and contentment in life. There is sex, fine food, and beautiful success, including all things foodie and gourmet. In 2020, the year of the Rat, it will be time to pay the bills after Pig year indulgence.

Wemoon born in Pig years (1911, 1923, 1935, 1947, 1959, 1971, 1983, 1995, 2007, 2019) are peaceful, optimistic and sensitive. A Pig wemoon has a heart of gold because she is a naturally kind, generous and gentle soul. Creative Pig can find success in the arts, including the culinary arts. In relationships, sincere Pig is faithful, devoted and affectionate. She is appreciated by her partner, who recognizes her pure heart. Pig is a romantic who makes sacrifices for those she loves. A Pig wemoon won't dazzle you like a Dragon or charm you like a Snake. Instead, she will grow on you until you cannot do without her. As a parent, Pig tends to spoil her children, and forgives all kinds of nonsense. Pig has high standards for herself, yet is rarely judgmental of others. At work, Pig is patient and dedicated. But this Pig year is not the time to attempt big or overwhelming career projects. Honest Pig can be naive, so be sure that no one takes advantage of you this year, especially in financial dealings. When threatened by change, Pig can become sick when experiencing too much turmoil. But do not mistake her kindness for weakness, Pig has amazing willpower.

Pig correlates to the Western sign Scorpio, but without the sting, because Pig is guileless. Pig is most compatible with another Pig, artistic Sheep and Rabbit, and brave Tiger.

Susan Levitt © Mother Tongue Ink 2018 www.susanlevitt.com

HERBS WITH SOLAR AFFINITY

The Golden Ratio © Lisa de St. Croix 2017

The Sun is the great nourisher, bringer of life, warming agent and catalyst for growth. And so are Her herbs. Now, more than ever, we need to work in harmony with the energy of the Sun in everything we do, from solar powered electricity to solar herbs. So, while you're dismantling the patriarchy and all its oppressive systems—racism, classism, capitalism, and misogyny—take time to warm up and sustain your activist body and soul. Here are some tips:

Some of the Sun's herbs look just like her. Calendula, *Calendula officinalis*, has a big sunny face and edible petals. The bright orange and yellow blooms add color and nutrition to your salads, and, infused in olive oil, make an incredible skin-healing salve. Sunflower, *Helianthus giganteus*, attentively follows that bright star in the sky throughout the day. Her head is a thousand tiny flowers arranged in sacred geometry. Sunflower seeds can be used to regulate your monthly cycle. Dandelion, *Taraxacum officinale*, those tenacious bits of sunshine in the yard, are rich in vitamin C, support the liver, can clear ringworm, and make a delicious salad.

Other solar herbs mimic the warming, illuminating properties of the Sun. Ginger, cayenne and cinnamon store up the heat from the Sun and chase the cold from our bodies.

Some solar herbs serve as tonics to the heart, your own radiant Sun housed in your solar plexus. Hawthorn, from the rose family, *Crataegus Spp.*, a heart tonic, energetically helps you stay open and protected. Motherwort, *Leonurus cardiac*, calms heart palpitations and restores a new mother's hormonal balance.

Sun herbs imbue vitality, health, creativity, dignity, success and authority. All herbs and plants capture their energy from the Sun. Through sunlight and photosynthesis, plants exhale the air we need to survive, and we do the same for them.

So, gratitude to the Sun for rising and to your heart for beating! Use solar herbs to strengthen and enliven your quest for a better world.

Sue Burns © Mother Tongue Ink 2018

THE SUN CARD IN TAROT

With Sun, we have reached the glorious potential of Creation. Here are fresh air and solid ground after navigating the deep waters of transiting Moon and the sometimes ungraspable brilliant Star. There is light and power here.

This card is about Trust. Trust in the fire that pushes us forward, and Trust that—when it feels like our coals are burning out—there

© OliviaJane Art 2015

are always embers glowing to heat our hearts. A reminder that the sun always shines, even on those who refuse to step outside their door and bask in the sunlight.

Nearing the end of Tarot's 21 card Major Arcana, these cards tend to bleed back and relate to the entirety of the Fool's journey which began with card 0. Sunlight cannot be contained, and its influence seeps and infuses all the cards in a spread of our lives. The essence of this card is outward movement. The Moon draws power in; the sun empowers us to act out, scream, sing and love loudly! Don't hold back; this card is telling us to speak our truth and fight for what we believe. Shine on!

Be careful though: this card can burn. At times, we need to cool down enough to attend to the voices of others. With generosity and understanding, we can use our Sun to give light and shed warmth on those who need it most. This card is an embodiment of abundance; in giving away, we receive our gifts back to ourselves tenfold. Embrace the inner sun child within; play with all the infinite possibilities that await in the sunshine. Remember to smile.

Tarot is a guide, a direct line to our instinct. It translates to our gut, insisting that we listen to what we already know. We never are surprised at a reading, rather reaffirmed and reestablished in ourselves. Tarot can guide the Muse; consult tarot when writing poetry and ask the cards to direct language on the page. Write unedited and unleash your inner Muse with the cards guiding your hand. Pull cards with no spread in mind. They will speak and act together if you open to them and trust yourself to see clearly.

Leah Markman © Mother Tongue Ink 2018

INTRO TO THE HOLY DAYS: HUMAN LUMINOUS

Visualize the Wheel of the Year this way: eight equidistant points on a circle, signifying the two solstices, the two equinoxes, and the midpoints between them. The points form the tips of an eight-pointed star, whose arms converge toward the center of the wheel. At center sits a round mirror. As the year revolves, the unfolding of the seasons is reflected in the woman gazing at the mirror—the woman we each are, at every turn of our own life. The seasons of the wheel mirror us.

Because the seasonal movements are created by the tilt of the Earth in relation to the Sun, the wheel is solar, one of changing, healing light. The turnings of the year not only reflect us; they season us. During the repeated turnings, our lived experience gradually extracts what is green; and we become seasoned wood, burning clean.

Enacting the Holy Days strengthens the bonds between our individual selves and the Gaia energy field of which we are a Divine aspect. The Wiccan narrative of the Wheel of the Year centers on the Goddess aging and rejuvenating with the seasons, and we match Her cycle to our own planet latitudes. On the Holy Days that mark the seasons, we honor Gaia and her changes, and we renew our relationship with Her. We look unflinchingly into the mirror of our own turnings, and how they affect our planet.

The light we see in the mirror at the heart of the wheel guides our evolution toward greater consciousness. Collectively, this opens the portal for our human evolution into the human luminous (aka *homo luminous*) of Mayan prophecy. As a butterfly emerging from its chrysalis is transformed, showing no trace of its previous caterpillar DNA, we can change human identity, by transforming ourselves, one woman at a time.

Wheel of the Year rituals accelerate personal and then planetary awakening; the resonance of our positive thought fields reaches toward the tipping point when a critical human mass awakens to our oneness. Living in the world in 2019 asks more of us as earth stewards than ever before. Celebrating Holy Day rituals helps season us into the luminous humans Gaia is crying out for us to become.

Susa Silvermarie © Mother Tongue Ink 2018

Wheel of the Year Chart © Gwen Davies 2014

RETURN OF THE SUN GODDESS

Radiance © Kay Kemp 2003

Once upon a time, when the Sun was revered as a Goddess, she was seen as one of the two equal eyes of heaven, embracing the world with radiant love, light and warmth by day. Her sister the Moon Goddess kept time with her by night, along with their Star sisters (ansisters) twinkling through the ages. And they, in turn, were surrounded by myriad goddesses who reflected the ever renewing source of life empowering every creature on Mother Earth. Together, they sustained the web of life in the natural balance each one upheld in her own way—for eons, since the beginning of time. As long as the Mothers' love was seen shining equally upon and within all their children, it is said that matriarchy held sway, and the warp and woof of humanity was inextricably woven into the natural matrix of harmony under the watchful eyes of the Mothers.

But sooner or later, the red thread of the Motherline began to lose hold, and a new story began to unfold. Some say it started a few thousand years ago, when people from the far North, where sunlight was scarce, came streaming southward proclaiming the Sun as God, the Father: the one and only ruler over all life on Earth. When it was eventually discovered (C.E.) that the Earth was round and circled around the Sun (instead of vice-versa), that clinched it: from then on, it was His story. The Sun came to be regarded as The Supreme Being at the center of the universe around which all else revolved. Ever since, believers in the One God have been at war with one another, depending on which religion claimed Him. In any case, men self-identified with Him in every day life on Earth, constructing themselves in His image to dominate the world. Henceforth, the evolution of human consciousness took an abrupt detour into the patriarchal paradigm. No longer was the Sun viewed as an Empowerer, a universal source of energy energizing the source of light shining within all beings; the Sun was considered more like an Emperor, ruling with an iron fist from above, with absolute power over all. Powerlessness became the dominant reality for most people, and women and nature were subjugated to the overbearing control of Man.

And so it happened that the Sun as Goddess came to be displaced. Women were disgraced, goddesses were banished, peoples and cultures that revered them were attacked and forcibly cut off from their ancestral roots. The wisdom traditions of indigenous peoples, and of pagans and witches who honor earth-based women's spirituality, have remained faithful to Her to this day. But the Great Mother Earth continues to be ruthlessly assaulted, exploited, and Her creatures threatened with extinction; Her inter-related web of life is being ripped apart by the rampages of domination, greed and violence.

As patriarchy spreads its virulent mix of perpetual warfare all over the earth, both within ourselves and among each other, the top-heavy world order we live in today is wobbling violently out of control, on the verge of collapse. The elemental terra-isms of wind storms, flooding, wild fires and drought are climaxing in the extremes of man-made climate change with hurricanes, tsunamis, earthquakes, and volcanoes—even more deadly than the rampant human terrorisms at large. In the balance of nature, the Earth Mother reigns.

Meanwhile, the self-serving Emperor has no clothes—and daily exposes himself for all to see. Humpty Dumpty sits atop his crumbling wall, poised to take a great fall. We may not know exactly how his-story ends or what comes next, but we can see light peeking through the deepening cracks in the wall, revealing the inevitable systemic collapse at hand. Even in these longest darkest nights, as the patriarchs fight hard to hold on, we see how out-of-control they actually are; they cannot hold back the dawn. A whole new world view is emerging with life-affirming consequences transforming all our relations. Women/wemoon, especially, are rising up now in acts of creative new leadership and resistance—bursting forth with power pent up since the Sun God first took over. We know from our foremothers that the Sun Goddess is returning now as She always does at the dawning of each new life cycle. Like broody hens, we are holding the eggs of the Mother World close under our wings, keeping them warm for as long as it takes. Can you feel them quivering now with new life?! And so Her-story begins again.

Musawa © Mother Tongue Ink 2018

INTRODUCTION TO THE THEME: FANNING THE FLAME

We'Moon 2019 is steeped in Sunglow! Our theme "Fanning the Flame" honors Mother Sun as the source of all life, and sparks our passion for revolutionary change in the world. She warms us and bestirs the molecules in every morsel of being. "Each wave, grain, scent, bee, is sunlight!" (Ann Filemyr, page 32). She also burns us, turns our forests and grasslands crisp, sucks up our waters—and so some of the mighty fanning we must do, on a massive scale, is to gather Her power, to use Her benign engine in place of the petrochemistry with which humans have poisoned Earth. Sun inspires us toward clarity and truth-speaking, impassions our longing for personal integrity, for just and life-affirming societies. Our fire ceremonies burn away fears, failures, occluded visions. We blaze with imagination, lift voice/drum/intention toward fresh expression and the Brand New, the Never Before. The holiness of Possibility.

Art and writing in *We'Moon 2019* speak reverence for Sun, and flame with urgency about protecting Earth, liberating creativity, restoring peace. Sun-inspired brightness—Lo! Even happiness—reverberates through poems about mothering and marching, art, gardens and ritual. Chapters entitled "WomanFire" and "Carrying the Torch" celebrate fierce work to blend Rage-at-injustice and Love-of-community. Change-making women dance at the ambiguous edge of cultural boundary, know that "The village matters now more than ever." (Emily Kedar, p. 143).

Our cover features Brigid, Goddess of Inspiration. Is She not a brilliant Sun Goddess? Is She not the quintessence of WomanFire? She is fierce to protect, determined In Her aim, focused on Her mission. She has gathered fire energy from Her ancient Irish origins, but not just: Her left hand forms the Agni Mudra or Fire Mudra, a ritual gesture in Hindu/Buddhist practice. She catalyzes the wide world, and throws Her blaze straight into the 21st century. Her anger is a sacred act of Spirit; Her heart, an inferno of renewal.

Here, She says to wemoon: Take this power of fire and flame it right into the face of modern devilry, ignite the end of patriarchy's ruling madness. STOP IT RIGHT NOW!

Here, wemoon: take this healing firelight into all the broken places; spread its wild and radiant Love. Our Sister Sun Goddess burns brightly. She clears the way for a New World.

Bethroot Gwynn © Mother Tongue Ink 2018

Amaterasu © Sandra Stanton 1996

Bright Mother Sun!

We thank you for Everything. You are our Everything.
We would be Nothing without your radiant light, your warmth
Your heat stored in earth-core, released in decay
Your photons powering the world's green
Your alchemy creating our life-breath
Your weight swinging us to and fro
but never letting loose of the orbital leash.
We are your awe-struck children.
Eos, Titan Goddess of Dawn—Bast, Lion Goddess of Sunset
You wake us to Beauty and Action, drowse us deep in creative fire
Amaterasu, Great Shining Heaven—Aditi, Source of All
Every second, 4 million tons of your mass explode into energy
We cry out wonder. We cry out fear.
Our shell is crisp and burning, our silver icing melts
We are pell-mell to protect our blue-green home
from the heavy blanket we wove, trapping your Dragon-fire.
Sweet Dragon, we ride your bright power into white-hot invention
Kindle the blaze of Vitality, Fan the flame of Spirit
Blow fervent with Enthusiasm, Insight, Regeneration
Gather our peoples around the HeartWarm altar, Sacred Sun,
where you are always smiling
And we are always Clear.

© Bethroot Gwynn 2017

Sun Worshipper
for Ondewéwé

You told me your great-grandmother
smoked her pipe and prayed to the sun.
Each morning she opened her window
and waited in the night for day to come.

It made me pause in my moon-gaze and wonder

For the next year all I thought about
was that single burning sky candle, that
fireball, shadow-thrower, white heart of dawn,
astonishing our world with a paint box of color.
In the great uncountable galactic realm,
only one star is ours.

Each wave, grain, scent, bee, is sunlight!
Even the copper moon,
rising and setting,
is night's sun.

We are starfire's brood:
blackberry, redhead, sunflower
gull feather, salt water, grandmother.

I became like your great-grandmother
turning my eye skyward,
watching that burning heart
love everyone-everything,
lathering the whole world
with golden ghee

Why wouldn't I worship
at the heals of that light
when all I love is made of sun?

0. IN PRAISE OF SUN

Moon 0: December 6–January 5

New Moon in ♐ Sagittarius Dec. 6; Sun in ♑ Capricorn Dec. 21; Full Moon in ♋ Cancer Dec. 22

Fire Ritual
© *Melanie Gendron 2015*

December 2018

desamber

))) Dŏsanbe

♈

Monday
17

) □ ♇ 7:20 am
☉ △) 6:27 pm
) ☌ ♅ 11:21 pm v/c

Sun Struck

You are my golden womb
the sphere I orbit
Days—I grow towards you
Nights—I dance in your reflection
You kindle me to life

excerpt © Joanne Rocky Delaplaine 2017

♂♂♂ Sešanbe

♈
♉

Tuesday
18

) → ♉ 1:37 am
) △ ♄ 7:27 pm
) ☍ ♀ 10:54 pm

☿☿☿ Cahâršanbe

♉

Wednesday
19

) ⚹ ♆ 2:33 am
) △ ♇ 1:40 pm
☉ □ ⚵ 2:02 pm
) ⚹ ♂ 4:41 pm v/c

♃♃♃ Panjšanbe

♉
♊

Thursday
20

) → ♊ 6:34 am
☉ △ ♅ 8:22 am
) ☍ ⚵ 9:41 pm
) ☍ ♃ 10:35 pm

♀♀♀ Jom'e

♊

Friday
21

) □ ♆ 5:58 am
♀ △ ♆ 9:11 am
☿ ☌ ♃ 9:37 am
☉ → ♑ 2:23 pm
) □ ♂ 9:40 pm

Winter Solstice

Sun in ♑ Capricorn 2:23 pm PST

ALL ASPECTS IN PACIFIC STANDARD TIME; ADD 3 HOURS FOR EST; ADD 8 HOURS FOR GMT

Sun Rays © *Jan Pellizzer 2017*

♊
♋

Saturday
22

☽⚹♅ 6:21 am v/c
☽→♋ 8:28 am
☉☍☽ 9:48 am

Full Moon in ♋ Cancer 9:48 am PST

♋

Sunday
23

☽☍♄ 1:17 am
☽△♆ 7:03 am
☽△♀ 9:55 am
☽☍♇ 5:23 pm

December 2018
diciembre

Lunar Awakening © Gaia Orion 2017

♋
♌

Monday
24

☽△♂ 12:37 am
☽PrG 1:46 am
☽□♅ 6:50 am v/c
☽→♌ 8:58 am
☿□♆ 4:32 pm

♌

Tuesday
25

☽△♃ 1:44 am
☽△♅ 9:06 am
☽□♀ 1:37 pm

♌
♍

Wednesday
26

☽△♅ 7:37 am v/c
☽→♍ 9:50 am
☉△☽ 6:33 pm

♍

Thursday
27

☽□♃ 3:48 am
☽△♄ 3:51 am
☽☍♆ 9:09 am
☽□♅ 4:04 pm
☽⚹♀ 6:51 pm
☽△♇ 8:04 pm

♍
♎

Friday
28

☽☍♂ 8:27 am v/c
☽→♎ 12:23 pm
♀⚹♇ 1:31 pm
♂♂♑ 9:41 pm

All aspects in Pacific Standard Time; add 3 hours for EST; add 8 hours for GMT

Every Nine Minutes

Scientists say it takes nine minutes
for the sun's brilliance to reach the earth.
Nine, the number of completeness,
of achievement.

Nine minutes seem too short
for many things: baking or making love.
Too long for others like goodbye
or a passing thought.
Waiting is always eternity.

I envision solar energy voyaging toward me:
passing Mercury with its scarred surface
and quick elliptical orbit. Next Venus,
Earth's cloudy sister, the hottest planet.
Light then gilding the dark side of the moon.

I open myself like a dawn flower.
I turn my check toward warmth.
I say: ignite my life with courage and commitment.
Every nine minutes, I am new.

© *Joanne M. Clarkson 2017*

─────── ♄♄♄ sábado ───────

♎︎

Saturday
29

☉□☽   1:34 am
☽□♄   7:41 am
☽⚹♃   8:00 am

Waning Half Moon in ♎ Libra 1:34 am PST

─────── ☉☉☉ domingo ───────

♎︎
♏︎

Sunday
30

☽□♇   12:23 am
☽⚹☿   2:13 am
☽⚹♅   2:53 pm v/c
☽→♏   5:23 pm

Dec. '18–Jan. '19

shí èr yuè / yī yuè

———— ▷▷▷ xīng qī yī ————

♏

Monday
31

⊙✳☽ 11:46 am
☽✳♄ 2:09 pm
♂→♈ 6:20 pm
☽△♆ 7:10 pm

———— ♂♂♂ xīng qī èr ————

♏

Tuesday
1

January

☽✳♇ 7:19 am
☽♂♀ 2:26 pm v/c
♄ApH 5:36 pm
⊙♂♄ 9:49 pm

———— ☿☿☿ xīng qī sān ————

♏
♐

Wednesday
2

☽→♐ 12:58 am
☽△♂ 2:41 am

———— ♃♃♃ xīng qī sì ————

♐

Thursday
3

☽♂♃ 12:23 am
☽□♆ 4:00 am
☿□♇ 2:41 pm
☿△♅ 9:13 pm

———— ♀♀♀ xīng qī wǔ ————

♐
♑

Friday
4

☽△♅ 8:10 am
☽♂☿ 9:41 am v/c
☽→♑ 10:55 am
⊙✳♆ 11:57 am
☽□♂ 4:05 pm
☿→♑ 7:40 pm

———

Lion Lotus © *Eileen M. Rose 2017*

♑ Saturday
5

♀△♂ 10:00 am
☽♂♄ 10:32 am
☽✶♆ 3:01 pm
☉♂☽ 5:28 pm
♀⊼♀ 6:28 pm

Partial Solar Eclipse 3:33 am PST*
New Moon in ♑ Capricorn 5:28 pm PST

♑
♒ Sunday
6

☽♂♇ 4:12 am
♅D 12:26 pm
☽□♅ 7:56 pm
☽✶♀ 10:20 pm v/c
☽→♒ 10:46 pm

MOON I - December 2018 / January 2019   *Eclipse visible in E. Asia and the Pacific   39

Walking
Paths of Fire

Like pyres, the black stacked
wood smokes chokingly,
until Lydia's shamanic drumming
ignites the flames to scarlet brilliance:
We are women returning to the Fire
still weeping for the Burning Times.
We take our priestess places in the heat;
in this wild night we'll walk on fire.
Sweat, tears and fears fall through us
as rain makes a spirit circle with the wind,
anointing grass with balm, the ember tracks
lie open to our nakedness, their hot red eyes
a gaze as fierce as lava.
We are stepping down an ancient path of courage
facing up to future, calling to our own power
to meet and hold this elemental force,
becoming one with it.

Beneath our feet burn solar forests. Our fears
are real enough to fill the sky with unlit darkness
for another thousand years but this night
there is a phoenix rising with a woman's face—
that it is hers, and yours, and mine—we feel
fire in our bodies, in our community of touch,
in our hearts thrown wide open to this challenge:
what is it that you fear, what is it you must face,
what fire will you walk for all of us, for Earth?

◻ *Rose Flint 2009*

I. CEREMONY

Moon I: January 5–February 4

New Moon in ♑ Capricorn Jan. 5; Sun in ♒ Aquarius Jan. 20; Full Moon in ♌ Leo Jan. 20

Fire Painter ¤ *Raven Bishop 2017*

January
Alel — Seaweed Month

───── ⟫⟫⟫ Henesháal — East Day ─────

≈

Monday
7

♀→♐ 3:18 am
☽✳♂ 7:42 am

───── ♂♂♂ Honesháal — West Day ─────

≈

Tuesday
8

☽✳♃ 1:44 am
☿□♂ 2:05 am
☽ApG 8:27 pm

───── ☿☿☿ Hunesháal — North Day ─────

≈
♓

Wednesday
9

☽✳♅ 8:53 am v/c
☽→♓ 11:44 am
☽□♀ 5:09 pm

───── ♃♃♃ Hanesháal — South Day ─────

♓

Thursday
10

☽✳♅ 4:16 am
☽✳♄ 1:13 pm
☽□♃ 3:48 pm
☽☌♆ 4:47 pm

───── ♀♀♀ Rayilesháal — Above Day ─────

♓

Friday
11

☉☌♇ 3:38 am
☽✳♇ 6:11 am
☉✳☽ 6:25 am v/c

Goddess of the Forest

She captures the wolf behind her cloak
they whisper through night
celebrate the birth
of winged ones and hooved
who gather in ritual
of new beginnings
along the river
of blackberry sage

Quiet the wind
as bird song
echoes tree roots
as hillsides question
sisters and brothers
of all species
who gather in honor
of night shades
and sunlight wisdom

Amber and Amethyst
resin and stone
return to the earth your light

© Marleine Rose 2017

Threshold of Light © *Dorrie Joy 2017*

ħħħ Yilesháal—Below Day

♓
♈

Saturday
12

☽→♈ 12:18 am
☽△♀ 11:20 am
☽♂♂ 4:12 pm

☉☉☉ Hathamesháal—Center Day

♈

Sunday
13

☽□☿ 1:05 am
☽□♄ 1:36 am
☽△♃ 4:31 am
♉♂♄ 5:31 am
♃□♆ 10:58 am
♇ApH 11:19 am
☽□♇ 5:28 pm
☉□☽ 10:45 pm

Waxing Half Moon in ♈ Aries 10:45 pm PST

January

zhanvīeh

Sea Otter ☐ *Molly Brown 2014*

◗◗◗ Dŏsanbe

♈
♉

Monday
14

☿✶♆	5:13 am
☽♂♅	7:56 am v/c
☽→♉	10:31 am

♂♂♂ Sešanbe

♉

Tuesday
15

☽△♄	10:50 am
☽✶♆	1:15 pm
☽△♅	5:29 pm

☿☿☿ Cahâršanbe

♉
♊

Wednesday
16

☽△♇	1:17 am
☉△☽	10:34 am v/c
☽→♊	5:00 pm

♃♃♃ Panjšanbe

♊

Thursday
17

☽☍♀	11:54 am
☽✶♂	12:32 pm
☽□♆	5:56 pm
☽☍♃	7:10 pm

♀♀♀ Jom'e

♊
♋

Friday
18

♀△♂	8:49 am
☿♂♇	12:03 pm
☉✶♄	4:57 pm
☉□♅	5:31 pm
☽✶♅	5:32 pm v/c
☽→♋	7:44 pm

ALL ASPECTS IN PACIFIC STANDARD TIME; ADD 3 HOURS FOR EST; ADD 8 HOURS FOR GMT

2019 Year at a Glance for ♒ Aquarius (Jan. 20–Feb. 18)

In 2019, your big ideas about the world find fertile ground in the company of others. If you already have a network of collaborators, draw on those relationships to manifest weird and brilliant plans. If you're more of a lone wolf Aquarius, then Jupiter encourages you to jump into the fray and join groups like neighborhood associations, political demonstrations, or philosophical societies. This year has great potential to provide fruition for your dreams, but to do this requires a shift in consciousness about how you connect with others.

You are in the midst of a deep lesson about compassion. While we humans have the neural infrastructure for empathy, we must be taught how to access it and use it. Saturn says it's time to overhaul your automatic responses to suffering and put compassion into daily action for yourself and others.

This year, Uranus in Taurus ignites a quest for your own roots. Seek out information about your own lineage, but also explore unusual subjects about how life and cultures survive. Find out which plants are most alive in the dead of winter. Learn about people who exist in extreme environments. Investigate the ways humans have coped in terrible circumstances—wars, persecution, natural disasters.

In July, the eclipses focus energy on your health. Routines may not be your thing, but you still have to care for your body on the regular. Mix it up by alternating modalities—a different one each week—sound healing, acupuncture, warm soaks, juicing, hydrating.

Rhea Wolf © Mother Tongue Ink 2018

We Were Seeds of Stars
© Emily Kell 2016

— ላላላ Šanbe —

♋

Saturday
19

☽□♂ 4:28 pm
☽⚹♄ 5:48 pm
☽△♆ 7:19 pm

— ☉☉☉ Yekšanbe —

♋
♌

Sunday
20

☉→♒ 12:59 am
☽⚹♇ 6:01 am
☽⚹♅ 10:56 am
☽□♅ 5:50 pm v/c

☽→♌ 7:54 pm
♀□♆ 8:15 pm
☉⚹☽ 9:16 pm

Sun in ♒ Aquarius 12:59 am PST
Total Lunar Eclipse 6:36 pm PST*
Full Moon in ♌ Leo 9:16 pm

January
enero

1991

Amaterasu: The Sun © Daughters of the Moon Tarot

—— ☽☽☽ lunes ——

♌

Monday
21

♂□♄	3:48 am
☽PrG	12:11 pm
☽△♂	6:19 pm
☽△♀	8:43 pm
☽△♃	9:11 pm

—— ♂♂♂ martes ——

♌
♍

Tuesday
22

♀♂♃	4:26 am
☿ApH	11:04 am
☽△♅	5:19 pm v/c
☽→♍	7:22 pm

—— ☿☿☿ miércoles ——

♍

Wednesday
23

☿□♅	3:13 am
☿✶♇	4:28 am
☽△♄	5:56 pm
☽☌♆	6:55 pm
☽□♃	9:41 pm
☿→♒	9:49 pm

—— ♃♃♃ jueves ——

♍
♎

Thursday
24

☽□♀	12:27 am
☽△♇	5:50 am v/c
☽→♎	8:02 pm
☽△♅	10:56 pm

—— ♀♀♀ viernes ——

♎

Friday
25

☉△☽	4:48 am
♂△♃	9:53 am
☽□♄	8:10 pm

The Properties of the Sun

In the Appalachian Mountains, it is believed that if you sweep your home after Sunset, you're doomed to live a life of poverty. They also have a legend that if the Sun shines during a rain storm, it will rain again at the same time the next day. A red sunrise means rain is coming soon.

Some Native American tribes practice the Sun Dance ceremony to honor the Sun as a manifestation of the Great Spirit and to bring the dancers visions. In some Wiccan and Pagan traditions, movement in

Sunset Walk © Denise Ostler 2017

the direction of the Sun (clockwise or deosil) is associated with gainful magic, and in the opposite direction (counter clockwise or widdershins) is connected to banishing or dismantling magic. In some magical traditions, workings performed during a solar eclipse are extra powerful—believed to obtain from the Sun greater vitality that increases your happiness levels, confidence, courage, charisma, and sense of freedom.

The Sun is also associated with numerous flowers and herbs: sunflowers, daisies, dandelions, chamomile, rosemary and more. In magic, you can use these plants to add an extra solar power.

excerpt ▢ Margarita Palma 2017

ħħħ sábado

♎︎
♏︎

Saturday
26

☽✶♃ 12:23 am
☽☌♂ 12:56 am
☽✶♀ 7:03 am

☽☐♇ 8:30 am
☽☍♅ 9:21 am v/c
☽→♏︎ 11:31 pm

☉☉☉ domingo

♏︎

Sunday
27

☽☐♉ 9:59 am
☉☐☽ 1:10 pm

Waning Half Moon in ♏︎ Scorpio 1:10 pm PST

January / February

yī yuè / èr yuè

♏

Monday
28

☽⚹♄ 1:46 am
☽△♆ 2:14 am
☽⚹♇ 2:39 pm v/c

Three of Fire

© Jan Kinney 1999

♏
♐

Tuesday
29

☽→♐ 6:33 am
☉♂☿ 6:52 pm

♐

Wednesday
30

☉⚹☽ 2:03 am
☽⚹☿ 2:32 am
☽□♆ 11:06 am
☽♂♃ 4:23 pm
☽△♂ 9:49 pm

♐
♑

Thursday
31

♄⚹♆ 6:15 am
☽♂♀ 9:35 am
☽△♅ 2:33 pm v/c
☽→♑ 4:47 pm

♑

Friday
1

February

♂□♇ 7:20 pm
☽⚹♆ 10:41 pm
☽♂♄ 10:57 pm

ALL ASPECTS IN PACIFIC STANDARD TIME; ADD 3 HOURS FOR EST; ADD 8 HOURS FOR GMT

Imbolc

Candlemas, a Sabbat of purification, initiation, and waxing light, celebrates the first signs of spring and marks the fertilizing warmth of the Goddess. Incoming energies are quickening everything! Outer light is increasing and infusing us as well, but the stretch to accommodate it within must be deliberate on our parts, so we can actively evolve toward human luminous. In this way, the ritual fire at Candlemas represents our own illumination as well as that of Sun Goddess Brigid.

Since the power shift away from the ancient matriarchal societies, female deities have been more associated with moonlight than sunlight, but we are once again re-membering the Sun Goddesses. Solntse, Pattini, Saule, Arinna, Uelanuhi, Aditi, Olwen, Amaterasu, Chaxiraxi the Sun Mother, and more. Call upon the Divine Face that fires up your spirit; your region, your culture.

When we ceremonially enact a seasonal change, we step Between, to an embryonic state when anything is possible. Pluripotency, the cell scientists call it. As the Imbolc song goes, "The seeds of life lie tingling" in the pluripotent moment when Winter promises to end. We shift! We step between realities and choose what our world will become.

Susa Silvermarie
© Mother Tongue Ink 2018

Brighid—Mother Goddess of Ireland © Jo Jayson 2012

Winter Ritual

The pile of sticks has grown all winter—
Sticks like witches' wands, weathered prayer sticks
picked-up sticks, built into a woody tent,
a grey and brown latticed cone.

Yesterday it snowed
and now I strike a match
to papers torn and pushed
under damp kindling.
I sit by the crackle,
am smudged
in silver smoke.

Butterfly Dream ◻ *Janis Dyck 2009*

I am burning the past
on its last remaining day:
Unsent letters
in unaddressed envelopes
Old journals,
pain pressed between pages
like dried flowers
Outdated bills, lists, memos
emptied from their files—
Time's linearity seared.

Flickers of orange curl back black edges,
thin and ephemeral as shed snake skin.
Words are lifted from their page,
carried off and transliterated
into the primal language of combustion.

Surrounding snow melt
trickles toward the wane of flames.
The day's final blaze flares
from mottled afternoon clouds behind me,
and warms my back
like a brief blessing.

Fanning the Flame © *Tamara Phillips 2017*

♑ Saturday

2

Imbolc / Candlemas

☽♂♇ 12:14 pm
☽□♂ 1:12 pm
♀△♅ 3:41 pm
♀□♄ 10:51 pm

♑
♒

Sunday

3

☽□♅ 2:53 am v/c
☽→♒ 5:03 am
☿⚹♃ 1:54 pm
♀→♑ 2:29 pm

Todo Tiene su Effecto © Adriana M. Garcia 2012

This mural celebrates the inter-connectivity of life. The cosmos reminds us of our ancestors' contributions. Past meets present with the brush-stroke of the muralists—stewards of our culture—as they paint stories of struggle, leadership and protest so generations can learn, and take ownership. The images come alive singing and playing music, dancing and laughing with friends. They celebrate our history, our present and our potential future. The face of the conscious individual moves forward, carrying experiences of her community. Her spirit meets inspiration represented by birds taking flight from her mind, guiding her towards a new day. She will create pathways to strengthen our connection with spirit, community and self.

II. FLAME OF CREATIVITY

Living Revolution

We often fear that the revolution needed is too big.
That we are too small.
But all that is required
is that you step into the truth of your life.
And speak it, write it, paint it, dance it.
That you shine your light on your truth,
for the world to see.
And as hundreds, then thousands,
then millions do this—
Each sparking the courage of yet more—
Suddenly, we have a world alight with truth.
We are shifting ourselves.
We are shifting the world.
Dancing her into a new orbit.
We are filling in the space
where our voices were silenced,
Filling in the blanks
where our images have been lacking.
We are weaving her-story into reality.
Reaching beyond his-story.
Into new ways of being and seeing.
We are the bridge between worlds.

© *Lucy H. Pearce 2016*

February

Ayáanin—Tree Month

─────))) Henesháal—East Day ─────

♒

Monday
4

Lunar Imbolc

☉☌☽ 1:03 pm
☽⚹♃ 6:35 pm
☽☌♅ 11:11 pm

───── ♂♂♂ Honesháal—West Day ─────

New Moon in ♒ Aquarius 1:03 pm PST

♒
♓

Tuesday
5

☽ApG 1:30 am
☽⚹♂ 5:49 am
☽⚹♅ 3:59 pm v/c
☽→♓ 6:02 pm
☽⚹♀ 11:33 pm

───── ♀♀♀ Hunesháal—North Day ─────

♓

Wednesday
6

No Exact Aspects

───── ♃♃♃ Hanesháal—South Day ─────

♓

Thursday
7

☽☌♆ 12:43 am
☽⚹♄ 1:44 am
☽□♃ 8:16 am
☽⚹♇ 2:13 pm v/c
☉⚹♃ 4:32 pm
☿⚹♂ 5:24 pm

───── ♀♀♀ Rayilesháal—Above Day ─────

♓
♈

Friday
8

☽→♈ 6:34 am
☽□♀ 6:21 pm

Seeking My Radiant Muse

Who is my true and lasting Muse?
Guide of my imagination,
Chorographer of my word,
Mystic whisperer?

Rain, what do you have to say?
Wind, what message do you bring?
Songbird, I am listening.

Violet and lily, is earth awakening my green?
Golden maple, what must I forgive and release?
Root, reach into my underground rivers.

Do clouds form my secret animal:
A deer, a mare, a feline?

Sunrise, open my veiled eyes.
Sunset, resolve the doubts of my day.

Grandmother, whose spirit has passed into rainbows,
Mirror sister with whom I comb each evening,
Which of you sparks all I can do and be and say?

I will dream you tonight.
I will call you into my dreams.

© Joanne M. Clarkson 2017

♄♄♄ Yilesháal—Below Day

♈

Saturday 9

☽□♄ 2:06 pm
☿⚹♅ 2:53 pm
☽△♃ 8:42 pm

☉☉☉ Hathamesháal—Center Day

♈
♉

Sunday 10

☉⚹☽ 12:39 am
☽□♇ 1:51 am
☿→♓ 2:50 am
☽☌♂ 12:48 pm
☽☌♅ 3:48 pm v/c
☽→♉ 5:28 pm
☽⚹☿ 7:57 pm

February
fevriēh

The prayer is just air
until you say it,
the heart is mute
until you sing it,
the ceremony is dormant
until you do it.
excerpt © Pam Ballingham 2009

―――― ☽☽☽ Dōsanbe ――――

♉ Monday
11

☽△♀ 10:39 am
☽⚹♆ 10:30 pm

―――― ♂♂♂ Sešanbe ――――

♉ Tuesday
12

☽△♄ 12:05 am
☽△♇ 10:54 am
☉□☽ 2:26 pm v/c
♂♂♅ 10:21 pm

―――― ☿☿☿ Cahâršanbe ―――― Waxing Half Moon in ♉ Taurus 2:26 pm PST

♉
♊ Wednesday
13

☽→♊ 1:32 am
☽□♅ 12:36 pm

―――― ♃♃♃ Panjšanbe ――――

♊ Thursday
14

♂→♉ 2:51 am
☽□♆ 4:56 am
☽☍♃ 12:56 pm
☉△☽ 11:48 pm

―――― ♀♀♀ Jom'e ――――

♊
♋ Friday
15

☽⚹♅ 4:48 am v/c
☽→♋ 6:03 am
☽⚹♂ 7:24 am
☽△♅ 11:39 pm

―――
ALL ASPECTS IN PACIFIC STANDARD TIME; ADD 3 HOURS FOR EST; ADD 8 HOURS FOR GMT

The Mezzo Goddess © *Emily Balivet 2008*

───── ꜩꜩꜩ Šanbe ─────

♋ ☾ ## Saturday
16

☽☌♀   6:23 am
☽△♆   7:48 am
☽☌♄   9:39 am
☽☌♇   6:39 pm

───── ☉☉☉ Yekšanbe ─────

♋
♌ ☾ ## Sunday
17

♀⚹♆   12:44 am
☽□♅   6:17 am v/c
☽→♌   7:21 am
☽□♂   10:57 am
☉⚹♅   11:55 pm

February
febrero

Monday
18

♋→♈ 1:10 am
♀♂♄ 2:52 am
☉→♓ 3:04 pm
☽△♃ 4:00 pm
♉♂♆ 10:37 pm

Sun in ♓ Pisces 3:04 pm PST

Tuesday
19

☽PrG 12:58 am
☽△♅ 5:51 am v/c
☽→♍ 6:47 am
☉☍☽ 7:53 am
☽△♂ 12:31 pm
☿✳♄ 6:39 pm

Full Moon in ♍ Virgo 7:53 am PST

Wednesday
20

☽☍♆ 7:22 am
☽△♄ 9:32 am
☽☍♉ 11:11 am
☽△♀ 1:44 pm
☽□♃ 3:41 pm
☽△♇ 5:52 pm v/c

Thursday
21

☽→♎ 6:17 am

Friday
22

☽□♄ 10:20 am
☿□♃ 12:40 pm
☽✳♃ 4:56 pm
☽□♀ 6:46 pm
☽□♇ 6:52 pm
♀♂♇ 7:53 pm

ALL ASPECTS IN PACIFIC STANDARD TIME; ADD 3 HOURS FOR EST; ADD 8 HOURS FOR GMT

2019 Year at a Glance for ♓ Pisces (Feb. 18–March 20)

What did you want to be when you grew up? The dreams of childhood can be a powerful catalyst for 2019. You don't have to dramatically shift course and become a ballerina-astronaut, but you can let the unlimited imagination of your child self inspire you as you make choices about your role in the world.

Opportunity calls you to step into more confidence in the outer world. There may be set-backs in the late spring, early summer, but you can use that time to reflect on what fills you with enthusiasm and passion. Jupiter may shift your consciousness about world-work, illuminating untrodden paths you're ready to walk with confidence.

This year, Saturn and Pluto demand that you think about boundaries in groups and organizations. Check in to see if you've been saying yes way too much, or if you've been shying away from responsibilities out of fear of doing it wrong. It might be time to shift your obligations as you learn about limitations from your community and friends. Sharing your hopes and dreams will help you hold onto optimism and resist power-over dynamics. Get out there and collaborate—without overfilling your schedule.

Uranus in Taurus offers innovation in simple, practical steps that transform your immediate environment—like a neighborhood school, community garden, or local newspaper. Roll up your sleeves, offer your skills, and get ready to participate in making surprisingly radical changes close to home.

Rhea Wolf © Mother Tongue Ink 2018

Halo © Susan Bolen 2016

ħħħ sábado

Saturday
23

☽ ☌ ♅   7:11 am v/c
☽→♏   7:56 am
☿⚹♇   9:18 am
☉△☽   4:44 pm
☽ ☍ ♂   7:12 pm

☉☉☉ domingo

Sunday
24

☽△♆   11:29 am
☽⚹ħ   2:21 pm
☽⚹♇   11:18 pm

February / March
èr yuè / sān yuè

────── ☽☽☽ xīng qī yī ──────

♏︎
♐︎

Monday
25

☽△♉ 3:14 am
☽⚹♀ 4:14 am v/c
☽→♐ 1:19 pm

The Dream Begins

────── ♂♂♂ xīng qī èr ──────

♐︎

Tuesday
26

☉□☽ 3:28 am
☽□♆ 7:15 pm

Waning Half Moon in ♐ Sagittarius 3:28 am PST

────── ☿☿☿ xīng qī sān ──────

♐︎
♑︎

Wednesday
27

☽♂♃ 6:33 am
☽□☿ 5:10 pm
☉⚹♂ 6:33 pm
☽△♅ 10:17 pm v/c
☽→♑ 10:48 pm

────── ♃♃♃ xīng qī sì ──────

♑︎

Thursday
28

☽△♂ 6:26 pm
☉⚹☽ 7:09 pm

────── ♀♀♀ xīng qī wǔ ──────

♑︎

Friday
1

March

♀□♅ 4:32 am
☽⚹♆ 6:39 am
♀→≈ 8:45 am
☽♂♄ 10:23 am
☽♂♇ 7:48 pm
♀⚹⚷ 9:55 pm

ALL ASPECTS IN PACIFIC STANDARD TIME; ADD 3 HOURS FOR EST; ADD 8 HOURS FOR GMT

© Stacie Haus 2016

Silence

Silence—
the bones on which
to hang my words,
my early morning treat.
The morning sun—
the kiln of inspiration,
the fuel to fire
my imagination's artist hand,
painting day's future perfect.

I bathe in it, the silence
and the sun—dress in it.
The grey cat regards me
with his yellow eyes
and nods approval.

Creator ¤ *Lisa J. Rough 2013*

Carefully, I unravel the silence
like strands of embroidery floss,
and make colourful patches on my dancing skirt.
I wrap a bit around my arrows called *Intent* and *Will*,
marking them as my own.
With whoops of joy
I loose them on my day.

¤ *The Obsidian Kat 2007*

ㅓㅓㅓ xīng qī liù

♑
♒

Saturday
2

☽⚹☿ 8:54 am
☽□♅ 10:47 am v/c
☽→♒ 11:06 am
☽☌♀ 2:03 pm

☉☉☉ lǐ bài rì

♒

Sunday
3

☽□♂ 10:53 am

Invocation to Happiness

Let happiness come
like snow melt
like creek swell
slipping down side of the mountain
gathering itself to itself

Let happiness come
moving toward you
as it has been for millennia
carving its way through stone

Let happiness come
to find you precisely where you are
Let it come
inevitable and surprising

Let happiness enter you
Let it overcome you
with taste and smell
like a new drink in your glass
each sip silver, sparkling
running down the channels of your body

Let happiness come
let it make your path smooth and sinuous
wider, deeper
let it come
and come and keep coming

then let it spread
like fingers of sweet water into the estuary
moving the rushes and grasses
flushing the salt that's waiting there.

III. SUNCHILD

Moon III: March 6–April 5

New Moon in ♓ Pisces March 6; Sun in ♈ Aries March 20; Full Moon in ♎ Libra March 20

Magical Child © Lindsay Carron 2017

Womb Words

Sweet heart, I don't know how to tell you this but you've swallowed the sky and the strange system over-culture has been crushed between your pearly shell-tipped toes. Something in the fierce love of your breath has tipped the balance of the world and now a full brilliance of species rejoice because Peace has wrapped them in her blanket. You and She are planning even now to set them joyously free under the righteous blessing of your sky-belly dawn.

excerpt © Nell Aurelia 2015

March
Ahesh—Grass Month

────))) Henesháal—East Day ────

≈

Monday
4

)ApG   3:27 am
)✶♃   8:30 am

──── ♂♂♂ Honesháal—West Day ────

≈
♓

Tuesday
5

)✶♅   12:05 am v/c
)→♓   12:11 am
☿R   10:19 am

──── ♀♀♀ Hunesháal—North Day ────

♓

Wednesday
6

♅→♉   12:26 am   )✶♄   12:58 pm
)✶♂   3:26 am   ☉♂♆   5:00 pm
☉♂)   8:04 am   )□♃   9:37 pm
)♂♆   8:47 am   )✶♇   9:41 pm

──── ♃♃♃ Hanesháal—South Day ────   New Moon in ♓ Pisces 8:04 am PST

♓
♈

Thursday
7

)♂♉   11:08 am v/c
)→♈   12:27 pm
♆ApH   12:50 pm

──── ♀♀♀ Rayilesháal—Above Day ────

♈

Friday
8

)✶♀   4:29 am
☉✶♄   11:10 pm

Guarded

Newborns smell like the inside of an exploding star. They smell like seedlings wiggling out of the ground. In the town of Nopales, we celebrate life and perpetual human demise. When someone from Nopales gives birth, they stop by the Mango House because Abuelita is considered a guardian for new souls. She ties thick strips of her white hair around chunky ankles and miniature wrists. The strips are to stay on their tiny bodies for a full thirty days. Abuelita does not claim divinity. She merely sits with new parents and grandparents and listens to their worries. Her hair is a corporeal gift that she gives them.

She listens to the women talk about their miraculous bodies. Some of them talk about being literally ripped in half...or of the bliss of hearing small lungs give way for oxygen and a beautiful wailing. Others claim to glimpse into the future, seeing their child go from fetal floating specimen into full-grown adult. Some have cried to Abuelita, because they aren't sure they can do it. She assures them, they can. They can bring a good human into being. There will be genuine pain. This is an inevitable truth. Being alive hands us cruelty at any given moment, but we tend to prevail. We are formations of skin and cells and somehow we survive.

Abuelita started the ritual as a symbolic gesture. One that says you have been listened to. There is someone here for you. This was true. She was open to phone calls and visits from anxious young mothers, or tired mothers, or depressed mothers. She was open to tell them, they are resilient bundles of cells.

excerpt from Itzá *published by Broken River Books © Rios de la Luz 2017*

ꜛꜛꜛ Yileshǎal—Below Day

♈
♉

Saturday
9

☽□♄ 12:44 am
☽□♇ 8:56 am
☽△♃ 9:14 am v/c
☽→♉ 11:10 pm
☽♂♅ 11:31 pm

☉☉ Hathameshǎal—Center Day

♉

Sunday
10

♂⚹♆ 9:20 am
☽□♀ 9:45 pm

Daylight Saving Time Begins 2:00 am PST

March
mars

Potatoes, kittens, children—
nurture, wonder, wait,
and celebrate.
□ *Lisa Kemmerer 2017*

———— ☽☽☽ Došanbe ————

☿

Monday
11

☽⚹♆ 7:17 am
☽♂♂ 8:26 am
☽△♄ 11:34 am
☉⚹☽ 4:13 pm
☽△♇ 7:13 pm

———— ♂♂♂ Sešanbe ————

☿
♊

Tuesday
12

☽⚹♅ 2:31 am v/c
☽→♊ 8:48 am

———— ☿☿☿ Cahâršanbe ————

♊

Wednesday
13

☉⚹♇ 7:29 am
☽△♀ 10:58 am
☽□♆ 2:43 pm
☉□♃ 6:29 pm

———— ♃♃♃ Panjšanbe ————

♊
♋

Thursday
14

☽☍♃ 2:50 am
♂△♄ 3:02 am
☉□☽ 3:27 am
☽□♅ 5:30 am v/c
☽→♋ 2:49 pm
☽⚹♅ 3:31 pm
☉♂♅ 6:47 pm

———— ♀♀♀ Jom'e ————

Waxing Half Moon in ♊ Gemini 3:27 am PDT

♋

Friday
15

♅□♃ 4:16 pm
☽△♆ 7:17 pm
☽☍♄ 11:24 pm

———————————————————

ALL ASPECTS IN PACIFIC DAYLIGHT TIME; ADD 3 HOURS FOR EDT; ADD 7 HOURS FOR GMT

Fool Moon Bloom © Lindy Kehoe 2017

♋
♌

Saturday
16

☽⚹♂ 1:22 am
☽☍♇ 5:53 am
☽△☿ 5:54 am
☿⚹♇ 6:08 am

☉△☽ 11:03 am v/c
☽→♌ 5:57 pm
☽□♅ 6:47 pm

♌

Sunday
17

☿⚹♂ 8:23 pm
☿PrH 11:50 pm

Spring Equinox

She comes, the Girrl Goddess, smart with heart, and girrl enough for the work of world healing. She communicates heart-to-heart, teaching us *ishin-denshin*, a Japanese expression for communicating through unspoken understanding, now a technology using the body for sound transmission.

The Sun sits vertically above the equator, magnanimously giving all of us equal day and night, whether vernal equinox in the Northern Hemisphere or autumnal equinox in the Southern. She dances us to equilibrium, balancing light and dark, reuniting Mother Goddess and Girrl Goddess.

This reunion within each woman and within the world births hope, which is no flimsy wish but is a ferocious stance—a conviction that our vision is fully possible: a planet without national boundaries where all the children of the world have food and clean water and shelter and love, where all are aware that we are co-creators with Gaia of the stunning earth. When we are done with dancing, we watch the flames and then the embers of our ritual sunfire, for further visions sent by the Mother.

Susa Silvermarie © Mother Tongue Ink 2018

Visual Lifesavers © Jennifer Smith 2008

Inner Child © Heather Rae Lawrence 2016

Freaking Brilliant

My small one wants to know—dreadfully, desperately—
When will it get easier? When will it stop hurting?
And those are not the questions, my love,
To heal you, to help you heal yourself, to help us heal our selves
The questions are: What can I learn here? What shall I create now?
What will I do in my day,
my moment upon moment deepening of life?
What will I do so I can live well with myself?

In my vision, in my dreamseed,
I am standing-still-rooted-running-dancing
rising-soaring-blossoming-shining
In a sunlit meadow
Drinking in, beaconing out,
The potent mix of all I can be, all I am

So, hold on, me:
Something freaking brilliant is about to emerge . . .

excerpt © Nell Aurelia 2015

March
marzo

Transformation © Kay Kemp 2017

))) lunes

♌
♍

Monday
18

D ☍ ♀   2:27 am
D □ ○   5:06 am
D △ ♃   8:19 am v/c
D → ♍   6:41 pm
D △ ♅   7:39 pm

♂♂♂ martes

♍

Tuesday
19

D PrG   12:57 pm
D ☍ ♆   9:13 pm

☿☿☿ miércoles

♍
♎

Wednesday
20

D △ ♄   1:15 am
D ☍ ♅   1:35 am
♂ △ ♇   4:41 am
D △ ♇   7:04 am
D △ ♂   7:10 am

☿ ✶ ♄   7:27 am
D □ ♃   8:22 am v/c
☉ → ♈   2:58 pm
D → ♎   6:28 pm
☉ ☍ D   6:43 pm

Spring Equinox

Sun in ♈ Aries 2:58 pm PDT
Full Moon in ♎ Libra 6:43 pm PDT

♃♃♃ jueves

♎

Thursday
21

♀ □ ♂   1:07 am
♀ ✶ ♃   7:16 am
♂ ⊼ ♃ 12:43 pm

♀♀♀ viernes

♎
♏

Friday
22

D □ ♄   1:39 am
D □ ♇   7:29 am
D ✶ ♃   8:59 am
D △ ♀ 11:10 am v/c
☉ ☌ ☿ 11:38 am
D → ♏   7:16 pm
D ☍ ♅   8:38 pm

ALL ASPECTS IN PACIFIC DAYLIGHT TIME; ADD 3 HOURS FOR EDT; ADD 7 HOURS FOR GMT

2019 Year at a Glance for ♈ Aries (March 20–April 20)

At many points in the last seven years, as Uranus moved through your sign, you probably wondered if you were having a stunning breakthrough or a nervous breakdown. But stress and anxiety are not the new normal, Aries. After one more flare up of activity and agitation in January and February you can get back to your baseline vitality—without the overwhelm. In the springtime, align with Uranus' move into Taurus by getting rid of what weighs you down, replenishing your body with green food or juices, and creating new priorities.

Saturn supports you in setting realistic, grounded goals to make your vocation more secure. The old mantra, "Work smarter, not harder," is a good guide as you revamp routines to conserve energy. Spending more time at home may not seem like a way to build positive career vibes, but you are currently learning how to protect your personal life, growing strong roots that support you. This may mean asking your ancestors for advice on how to engage in the world or setting up a sanctuary in your home to make down time more nourishing.

Jupiter's romp in Sagittarius will prompt some exploration into your philosophy or spiritual views. Enter a program of study, travel, or join a discussion to find new meaning in life. You were born to *do*, Aries. But your shift in consciousness depends on finding out what it's like to *be*—and how your value is not based on action alone.

Tidal Bloom
© *Tamara Phillips 2014*

Rhea Wolf © Mother Tongue Ink 2018

ħħħ sábado

♏ ☽ **Saturday**
23

☽△♆ 11:50 pm

☉☉☉ domingo

♏ ☽ **Sunday**
♐   **24**

☽△⚳ 12:10 am   ☽⚹♇ 10:30 am
☽⚹♄   4:25 am   ☽☌♂   3:37 pm
⚥☌♆ 10:29 am   ☽□♀   7:24 pm v/c
⚵ApH 10:30 am   ☽→♐ 11:06 pm

March

sān yuè

♐

Monday
25

☉△☽ 7:30 am

Lingering Trees □ Janis Dyck 2014

──── ♂♂♂ xīng qī èr ────

♐

Tuesday
26

☽□♅ 5:01 am
☽□♆ 6:07 am
♀→♓ 12:43 pm
☽♂♃ 7:37 pm v/c

──── ☿☿☿ xīng qī sān ────

♐
♑

Wednesday
27

☽→♑ 7:07 am
☽✳♀ 9:06 am
☽△♅ 9:10 am
♀✳♅ 9:45 am
☉□☽ 9:10 pm

Waning Half Moon in ♑ Capricorn 9:10 pm PDT

──── ♃♃♃ xīng qī sì ────

♑

Thursday
28

♅D 6:59 am
☽✳♅ 2:48 pm
☽✳♆ 4:33 pm
☽♂♄ 10:00 pm

──── ♀♀♀ xīng qī wǔ ────

♑
♒

Friday
29

☽♂♇ 4:35 am
☽△♂ 5:05 pm v/c
☽→♒ 6:46 pm
☽□♅ 9:10 pm

ALL ASPECTS IN PACIFIC DAYLIGHT TIME; ADD 3 HOURS FOR EDT; ADD 7 HOURS FOR GMT

Regeneration

In the world we create, our daughters wield power so mighty it blinds the dark past. They show us their swords, stirring wind into flame.

With their strength, the trees are protected. Acres of jungle, forests of green, all creature homes, now intact. Clean air and water, above and below, our daughters leading the pack.

¤ Angela Bigler 2017

The Huntresses. Reflection
© Paula Franco 2017

ᚼᚼᚼ xīng qī liù

Saturday
30

⊙✶☽ 2:53 pm
♂→♊ 11:12 pm

⊙⊙⊙ lǐ bài rì

Sunday
31

☽ApG 5:10 pm
☽✶♃ 8:01 pm v/c

April

Athil—Vine Month

The Cosmic Walk Home
My children will remember my wisdom
when they look to the stars,
for that is from where I came
and that is where I return.
excerpt © Samantha Carney 2017

———— ᒧᒧᒧ Henesháal—East Day ————

♒
♓

Monday
1

☽→♓ 7:48 am
☽□♂ 9:44 am
☽✶♅ 10:29 am
☽☌♀ 11:31 pm

———— ♂♂♂ Honesháal—West Day ————

♓

Tuesday
2

☿☌♆ 2:35 am
☽☌♆ 6:25 pm
☽☌☿ 6:58 pm
☽✶♄ 11:58 pm

———— ☿☿☿ Hunesháal—North Day ————

♓
♈

Wednesday
3

☽✶♇ 6:09 am
☽□♃ 8:36 am v/c
☽→♈ 7:56 pm
♂✶⚷ 8:39 pm

———— ♃♃♃ Hanesháal—South Day ————

♈

Thursday
4

☽✶♂ 1:16 am

———— ♀♀♀ Rayilesháal—Above Day ————

♈

Friday
5

☉☌☽ 1:50 am
☽□♄ 11:01 am
☽□♇ 4:51 pm
☽△♃ 7:15 pm v/c

New Moon in ♈ Aries 1:50 am PDT

Illuminata

Oh Holy Matter who brings us out of darkness and into being, we are ripples of energy moving on the wave of life. Sparks jumping to be fanned into flames that ignite our existence. The light nurtures us as atoms collide, molecules swirl, and DNA twirls us into something miraculous. Through millions of millennia synapses fire as ancient embers ignite memories, and the dejá vu of our eternity arises again, and again, and again.

We are your sacrament of water, air and dust—of worlds built and consumed by wind, and oceans, and time. Our soft earthly temples hold altars to the energetic essence of our being as we wax and wane in your timeless trinity of life, death, rebirth. We are gifted with the dust of passing stars, bestowing upon us the cosmic mystery of your gifts. You bring us light and sight, and visions of all that we were, we are, and will be. And in the coming and going, in the void and fullness, in the dark and the light, there you are—Sacred Divine Eternal Matter.

© Marty Hamed 2017

ካካካ Yileshául—Below Day

♈
♉

Saturday
6

☽→♉   6:06 am
☽☌♅   9:09 am

☉☉☉ Hathameshául—Center Day

♉

Sunday
7

☿⚹♄   2:17 am
☽⚹♀   9:04 am
☽⚹♆   2:46 pm
☽△♄   7:59 pm
☽⚹♅   9:04 pm

Rising

can we stretch to accommodate new understanding; unfold to new and deeper form, in the sweet-reach challenge dance of evolving, of discovery; drop the veils layer by layer until all we see in the mirror is transformation?

i'm calling to be stream-lined forge-fire-refined, shed of old fears and patterns, outmoded survival strategies and wound deflectors. i want to tend the ugly as much as the lovely, and, all along, know that we two are equals: frail-strong light-and-shadow humans finding their way . . . no one is in the wrong. no one is in need of forgiveness from the other. because, to me, that's the pact: we are human, we are guardians of the richness of our souls and where they will take us in their endless desire to learn.

we don't need to hide these vulnerable, confused, hurting parts from each other, pretend that they are "other" or someone else is to blame. we are equal in life and love, in confusion and resolution we are equal. defenses are not needed nor pretences either.

so, one soul to another, let's grow—tend the heart, quiet the mind, acknowledge the emotions, and above and below and within all that, hear the soul's calling: to understand, to learn, to grow, to

know we are love—originated, designated, fibre being, spirit dreaming love. it's what i'm made of. certain. like the rest of creation. the spirit is clear and so are the stars; there's nothing to fight here. only allow what is true, return to our original fierce freedom of being love. no fall   no find   no search no loss   no fear   no looking no grabbing—open and rising to receive the true love of all we are.
© Nell Aurelia 2015

Love Thy Neighbor
© Francene Hart 2012

IV. LOVE SPARKS

Moon IV: April 5–May 4

New Moon in ♈ Aries April 5 ; Full Moon in ♎ Libra April 19; Sun in ♉ Taurus April 20

Dream Electric Dance #2 ▫ *Gretchen Butler 2015*

April

avrīl

☽☽☽ Došanbe

♉ ♊

Monday 8

☽△♇ 1:29 am v/c
☽→♊ 2:15 pm

New Beginnings

© Jodi Reeb 2008

♂♂♂ Sešanbe

♊

Tuesday 9

☽☌♂ 1:15 am
☽□♀ 9:50 pm
☽□♆ 9:58 pm
♀☌♆ 11:13 pm

☿☿☿ Cahâršanbe

♊ ♋

Wednesday 10

☉□♄ 1:47 am
☉✶☽ 3:06 am
☽□☿ 7:43 am
♃R 10:01 am
☽☍♃ 10:27 am v/c
☿✶♇ 2:45 pm
☽→♋ 8:31 pm
☽✶♅ 11:49 pm

♃♃♃ Panjšanbe

♋

Thursday 11

☿□♃ 9:18 pm

♀♀♀ Jom'e

♋

Friday 12

☽△♆ 3:14 am
☽△♀ 8:05 am
☽☍♄ 8:05 am
♀✶♄ 8:08 am
☉□☽ 12:06 pm
☽☍♇ 1:01 pm
☽△☿ 4:33 pm v/c

Waxing Half Moon in ♋ Cancer 12:06 pm PDT

ALL ASPECTS IN PACIFIC DAYLIGHT TIME; ADD 3 HOURS FOR EDT; ADD 7 HOURS FOR GMT

Pear

In a maple and beech forest
in the stairwell of a girls' dorm, in the kitchen of an old hotel
on a black rotary phone with a twisted cord
on the faded living room carpet of an old bungalow
on the forgotten sand dune of a melted glacier
in an abandoned farmhouse beside the Grand River
in a garden of broken abalone, in a burned down button factory
in salt spray, in a rain forest full of rain
near the bend in the river, near an altar to Kali
near railroad tracks on the Westside
near a bathtub of Israeli salt
in an old sweater with frayed elbows, in a red silk slip
in bare feet, on the Moray Firth
on the fourth floor, on the fire escape
in an office with a single window
behind a closed door, behind a dead elm
behind a stone wall, in the window of a closed café
in the doorway of a Hindu temple
in that flat fertile stretch of drained swamp
in her pear hips and her pear belly
with a pear in her hand and a pear in her mouth
beside the old willow, where fear meets longing
she touches that terrible mysterious thing
some call love

© Ann Filemyr 2011

ᎭᎭᎭ Šanbe

ᏋᏂ
Ꮆ

Saturday
13

D→Ꮍ 12:50 am
⊙□ᴾ 1:07 am
D□ᵂ 4:13 am
D⚹♂ 4:13 pm

⊙◉◎ Yekšanbe

Ꮍ

Sunday
14

⊙△♃ 6:40 am
♀⚹ᴾ 4:51 pm
D△♃ 5:49 pm
⊙△D 6:38 pm v/c

April
abril

© Sudie Rakusin 1991

♌
♍

Monday
15

☽→♍ 3:14 am
☽△♅ 6:42 am
♀□♃ 4:15 pm
☽□♂ 8:28 pm

♍

Tuesday
16

☽☌♆ 8:09 am
☽△♄ 12:40 pm
☽PrG 3:13 pm
☽△♇ 5:10 pm
☽□♃ 7:03 pm
☽☍♀ 9:29 pm v/c
☿→♈ 11:00 pm

♍
♎

Wednesday
17

☽→♎ 4:22 am
☽☍♅ 4:51 am
☽△♂ 11:46 pm

♎

Thursday
18

☽□♄ 1:48 pm
☽□♇ 6:18 pm
☽✳♃ 8:06 pm

♎
♏

Friday
19

☉☍☽ 4:12 am v/c
☽→♏ 5:40 am
☽☍♅ 9:35 am
☿☌♅ 4:47 pm

Full Moon in ♎ Libra 4:12 am PDT

ALL ASPECTS IN PACIFIC DAYLIGHT TIME; ADD 3 HOURS FOR EDT; ADD 7 HOURS FOR GMT

2019 Year at a Glance for ♉ Taurus (April 20–May 21)

Taurus—you dependable, stubborn beauty—it's time to unleash the ox from its yoke and let the beast go wild! Uranus' move into your sign unleashes a surge of unexpected energy throughout 2019. If you ignore the signs, this year could knock you for a loop. By learning how to ride the waves of chaos, you'll transform discord into synchronicity.

This year, unusual people may enter your life, or you may try on a new identity. Take in the messages without getting hooked on the messengers. Your Achilles' heel is attachment, but the universe wants you to keep changing for the next few years. These unusual relationships and strange inclinations are here to burn away stuck places inside of you. At the same time, it will be important for you to hold onto spiritual practices that make you feel secure. Saturn suggests that daily devotions or traditional prayers will keep you grounded even as pandemonium reigns.

Jupiter provokes a shift in consciousness by revealing hidden places deep inside of you—including painful wounds from the past. Release any energy bound up in grudges or resentment. Opportunities to speak your truth, even the truth that scares you, will occur with the eclipses in January and July. During the summer months, practice empathic responses to situations. Consider joining a support group, healing your sexual side, or bringing more mysticism into your life. Your revolutionary heart ignites when you get out of your comfort zone.

Rhea Wolf © Mother Tongue Ink 2018

———— ħħħ sábado ————

♏ ○   Saturday
         20

☉→♉   1:55 am
♀→♈   9:10 am
☽△♥ 11:39 am
☽⚹♄   4:20 pm
☽⚹♇   9:00 pm v/c

Sun in ♉ Taurus 1:55 am PDT

———— ☉☉ domingo ————

♏ ◐   Sunday
♐   21

☽→♐   8:59 am
☽△♀ 11:19 am
☽△♥   8:18 pm

April
sì yuè

Flaming Love © Cathy McClelland 2017

──────── ☽☽☽ xīng qī yī ────────

♐

Monday
22

☽☌☌ 11:35 am
☉☌♅ 4:07 pm
☽□♆ 5:03 pm

──────── ♂♂♂ xīng qī èr ────────

♐
♑

Tuesday
23

☽☌♃ 4:43 am v/c
♅ApH 6:01 am
♀☌♇ 9:55 am
☽→♑ 3:50 pm
☽△♅ 8:46 pm
☉△☽ 11:02 pm

──────── ☿☿☿ xīng qī sān ────────

♑

Wednesday
24

☽□♀ 12:12 am
☽□♅ 11:14 am
♇R 11:48 am

──────── ♃♃♃ xīng qī sì ────────

♑

Thursday
25

☽⚹♆ 2:22 am
☽☌♄ 7:33 am
☽☌♇ 12:48 pm v/c

──────── ♀♀♀ xīng qī wǔ ────────

♑
♒

Friday
26

☽→♒ 2:27 am
☽□♅ 7:57 am
☉□☽ 3:18 pm
☽⚹♀ 5:58 pm

Waning Half Moon in ♒ Aquarius 3:18 pm PDT

ALL ASPECTS IN PACIFIC DAYLIGHT TIME; ADD 3 HOURS FOR EDT; ADD 7 HOURS FOR GMT

Illumination

I shall risk the beauty
of delicate wine glasses
which may, indeed, break.
For it is my own beauty
I am after.

I shall risk the bonding
that fractures my heart
and summons old stories.
For it is my own love
I am after.

I shall leap from the raft
I misperceived as safety,
and risk the sea of love.
For it is my own deeps
I seek.

I shall attend the sunsets
showing up devoted and solo
with my flaming tickets,
for it is my own radiance
I find.

¤ *Susa Silvermarie 2017*

Seated in Sun ¤ *Arna Baartz 2016*

ちちち xīng qī liù

Saturday
27

♂□♆ 6:03 am
☽✶☿ 7:35 am
☽△♂ 3:10 pm

☉☉☉ lǐ bài rì

Sunday
28

☽✶♃ 2:44 am v/c
☽ApG 11:25 am
☽→⧓ 3:11 pm
☽✶♅ 9:02 pm

April / May

Athil—Vine Month / Amahin—Flower Month

A Dreamy Days Night
© Serena Supplee 2012

))) Henesháal—East Day ——————

♓ **Monday**
29

☉∗☽ 9:34 am
♄R 5:54 pm

♂♂♂ Honesháal—West Day ——————

♓ **Tuesday**
30

☽♂♆ 3:33 am
☽□♂ 7:22 am
☽∗♄ 8:34 am
☽∗♇ 1:48 pm
☽□♃ 2:57 pm v/c
☿∗♂ 11:37 am

☿☿☿ Hunesháal—North Day ——————

♓
♈ **Wednesday**
1

☿□♄ 1:50 am
☽→♈ 3:24 am
♂⊼♄ 5:17 am

May
May Day / Beltane

♃♃♃ Hanesháal—South Day ——————

♈ **Thursday**
2

☽♂♀ 7:39 am
☿□♇ 2:51 pm
☽□♄ 7:16 pm
☿△♃ 8:59 pm
☽∗♂ 9:22 pm

♀♀♀ Rayilesháal—Above Day ——————

♈
♉ **Friday**
3

☽□♇ 12:17 am
☽△♃ 1:06 am
☽♂♉ 1:47 am v/c
☽→♉ 1:18 pm
☽♂♅ 7:15 pm

ALL ASPECTS IN PACIFIC DAYLIGHT TIME; ADD 3 HOURS FOR EDT; ADD 7 HOURS FOR GMT

Beltane

We reclaim the original meaning of Virgin, a woman who belongs only to herself. We look in the mirror at the center of the wheel and tell ourselves: I am a woman faithful to herself. Call me no name, nor wife, nor mother, call me Innocent-of-Name. This virginity, never lost, is our indelible nature.

Beltane celebrates safe passage through the dark into the exploding passion and fertility of plants and animals. We light our sunfire as the sign of lighting the fire of our sexual power, giving special thanks for that part of our female body with the single function of pleasure. The Charge of the Goddess reminds us: All acts of love and pleasure are my rituals. Whether savoring solitude as a party of one, trysting with another, or whirling in community round a Maypole, we call forth the memory of the Temple Virgin, conduit to the Great Cosmic Mother.

As channels, our power to transform the world burgeons. We sing! To the trees, the water, the winds. We sing fierce incantations to heal our Gaia! We dance and love with every fiber of our beings— until we drop like children into Her bountiful lap.

Susa Silvermarie © Mother Tongue Ink 2018

Spring Blooms © Catherine Molland 2013

The Fool

The Fool has followed me home,
in her coat of fresh delight—
harlequin.
A cheerful choir of rainbow diamonds,
a thousand buttons close it.
She's sitting on the settee
opposite me, next to the dried sunflowers
that need flicking with a duster, or throwing out,
her happy little dog is grinning by her ankles.
She says: "I was born free,
on a beach, with giant turtles,
it was a very humbling experience."
I snigger self consciously, I haven't got time for this,
I've got the 'to do' list to do, the bills to pay,
the hamster to take to market.

She winks,
her eyes are greener than the forest floor in spring,
they turn aqua blue, chocolate, vermillion
she reaches down and opens her first button,
every button hole is a spy hole,
on a land she's danced across,
cliffs of change she's blithely leapt off
lifting up the clouds closer to the sun
with her great gasps of pleasure.
I can only stare,
it's the land of primal fire
when life was just a fish thought
when we were soup
before we got our eyes,
our fingerprints,
our souls,
our longing.

◻ *Debra Hall 2014*

She Awakens and Gathers Her Powers © Denise Kester 2005

ꝭꝭꝭ Yilesháal—Below Day

 ☿

Saturday
4

☉☌☽   3:45 pm
☽⚹♆   11:02 pm

New Moon in ♉ Taurus 3:45 pm PDT

◉◉◉ Hathamesháal—Center Day

 ☿
♊

Sunday
5

☽△♄   3:22 am
♂⚺♇   5:29 am
☽△♇   8:10 am v/c
♂☍♃   2:57 pm
☽→♊   8:40 pm

Dawn Prayer Woman

In the stillness of not-yet-light, dog-woken, I rise from sleep,
hurrying into mis-matched clothes, fumbling with door-locks,
stepping out into anticipation—
Sky-ward seeking eyes scan the horizon, open wide, receive deep.
Humming myself into alignment with Mama Earth, I wait . . .

On the best of days
the Sunrise breaks me open

Open—quiet inbreath of astonishment and recognition
Open to my deepening Self,
Open to that Place Between.

In a glimpse of clarity,
seeing the new/old rede of the new/old day,
in the same spirit as my blood-mothers
of many places and many times,
I sing my soul to the newest Old Sun,
chanting my libation to Mama Earth.

¤ *Lissa Callirhoe 2017*

Sunwashed ¤ *Liz Darling 2016*

V. ENLIGHTENMENT

Moon V: May 4–June 3

New Moon in ♉ Taurus May 4; Full Moon in ♏ Scorpio May 18; Sun in ♊ Gemini May 21

Sun Salutation © *Jenny Hahn 2004*

May
meh

☽☽☽ Došanbe

♊

Monday
6

☿→♉ 11:25 am

♂♂♂ Sešanbe

♊

Tuesday
7

☽□♆ 5:15 am
♀□♄ 6:26 am
☽✶♀ 9:35 am
☽☍♃ 2:10 pm
☽☌♂ 4:50 pm v/c

☿☿☿ Cahâršanbe

♊
♋

Wednesday
8

☽→♋ 2:06 am
☿☌♅ 7:22 am
☽✶♅ 8:06 am
☽✶☿ 8:13 am
☉✶♆ 7:52 pm

♃♃♃ Panjšanbe

♋

Thursday
9

☽△♆ 9:55 am
♀△♃ 9:56 am
♀□♇ 10:20 am
☉✶☽ 10:57 am
☽☍♄ 1:44 pm
☽☍♇ 6:20 pm
☽□♀ 7:06 pm v/c

♀♀♀ Jom'e

♋
♌

Friday
10

☽→♌ 6:14 am
☽□♅ 12:19 pm
☽□☿ 8:33 pm

All aspects in Pacific Daylight Time; add 3 hours for EDT; add 7 hours for GMT

The Leaning Intention

There are those exotic foreign words
Those phrases that cannot be
That defy translation because
Another language doesn't know
How to invite them in
Is ill equipped to receive them
(Where should we house them?
In the attic? Guest room or garage?
And how long are they planning to stay?)
Words that break the translation mold
Gears smoking, gogs clogged, equipage mute and smouldering
As though an alien world had lassoed the tongue
There are such places in the heart as well, as remote, as untranslatable
High above the timber line of word pines where the air is so thin
Language has yet to find a mouth, let alone an alphabet
Where words are no more than a suggestion
A leaning intention
A hovering cloud mixed with a hint of rain
(Tethered hearts ascending like balloons graduating from the sky)
The notion before the invention
The shadow of flames
Dancing on the walls of the cave

Sun Dream
© Gretchen Butler 2015

¤ Shelley Blooms 2017

— ʜʜʜ Šanbe —

♌ Saturday

11

⊙△♄ 2:19 am
⊙□☽ 6:12 pm
☽△♃ 9:16 pm

Waxing Half Moon in ♌ Leo 6:12 pm PDT

— ⊙⊙⊙ Yekšanbe —

♌
♍ Sunday

12

☽△♀ 3:14 am
☽⚹♂ 5:24 am v/c
☽→♍ 9:22 am
☽△♅ 3:33 pm

May
mayo

Monday
13

☽△♅ 7:48 am
☉⊼♃ 11:27 am
☽PrG 2:51 pm
☽☍♆ 4:14 pm
☽△♄ 7:39 pm
☉△♇ 8:07 pm
☽□♃ 11:32 pm

Web Weaver

Tuesday
14

☽△♇ 12:12 am
☉△☽ 12:29 am
♀⚹♂ 6:58 am
☽□♂ 10:19 am v/c
☽→♎ 11:51 am

Wednesday
15

♀→♉ 2:46 am
☿⚹♆ 6:20 pm
♂→♋ 8:09 pm
☽□♄ 10:00 pm

Thursday
16

☽⚹♃ 1:39 am
☽□♇ 2:37 am v/c
☽→♏ 2:26 pm
☽△♂ 3:18 pm
☿△♄ 4:09 pm
☽☍♀ 5:47 pm
☽☍♅ 9:03 pm

Friday
17

☿⊼♃ 2:49 pm
☽△♆ 10:02 pm
☿△♇ 10:48 pm

Psalm: Admiring a Spider's Work
High in the Branches of the Peach Tree at Sunrise

strands spun so slender they are imaginary
     until made real
         by light

flashes of silver
move like a melody
a humming
between the green thickness of leaves

the wonder is the way that each reveals the other,
     without the strand,
     no sparkling wire of light
     without the light,
     no gleaming silver strand

which is it
     particle or wave?
reality sparks forth
leaping out from the flinty blades of duality
delightful
wondrous
something we have no need to understand

© *Cathy Casper 2014*

ᚺᚺᚺ sábado

♏︎   ( )   Saturday   Lunar Beltane
♐︎   18

☽✷♄   1:14 am   ♀☌♅   9:17 am
☽✷♇   6:04 am   ☉☍☽   2:11 pm v/c
☽☍♅   7:26 am   ☽→♐   6:21 pm

Full Moon in ♏︎ Scorpio 2:11 pm PDT

☉☉☉ domingo

♐︎   ( )   Sunday
               19

No Exact Aspects

May
wŭ yuè

———— ♪♪♪ xīng qī yī ————

♐

Monday
20

☿ApH 2:13 am
☽□♆ 3:32 am
☽♂♃ 10:05 am v/c

Limbic Resonance © Amanda Sage 2011

———— ♂♂♂ xīng qī èr ————

♐
♑

Tuesday
21

☽→♑ 12:56 am
☉→♊ 12:59 am
☿→♊ 3:52 am
☉♂☿ 6:07 am
☽☍♂ 7:35 am
☽△♅ 8:43 am
☽△♀ 3:58 pm

Sun in ♊ Gemini 12:59 am PDT

———— ☿☿☿ xīng qī sān ————

♑

Wednesday
22

♂⚹♅ 7:46 am
☽⚹♆ 12:14 pm
☽♂♄ 3:22 pm
☽♂♇ 8:58 pm v/c

———— ♃♃♃ xīng qī sì ————

♑
♒

Thursday
23

☽→♒ 10:49 am
☿⚹♇ 11:04 am
☉△☽ 3:49 pm
♂□♇ 4:11 pm
☽□♅ 7:17 pm
☽△☿ 11:01 pm

———— ♀♀♀ xīng qī wŭ ————

♒

Friday
24

☽□♀ 9:17 am

ALL ASPECTS IN PACIFIC DAYLIGHT TIME; ADD 3 HOURS FOR EDT; ADD 7 HOURS FOR GMT

2019 Year at a Glance for ♊ Gemini (May 21–June 21)

Your natural curiosity makes you quite comfortable dealing with the inexplicable. But to understand the revolutionary potential of this year, you will have to learn patience, quiet reflection, and planning—difficult tools for you to wield.

Uranus will provide opportunities to innovate your spiritual life radically. But Gemini, you can't think your way out of a spiritual dilemma. Your body, and the spiritual wisdom contained therein, is begging for your attention. Try on new modalities such as herbalism, body-mind practices, or gardening-meditation.

Pluto and Saturn stir up deep emotions to teach you how to face your fears. Don't run away from these hard feelings! If you can sit with the discomfort of vulnerability, you will integrate new resources of power. Harness this Capricornian motivation to see the sacred in the taboo, and create space for challenging community conversations.

Jupiter brings a shift in consciousness as close friends radiate inspiration and guidance. You may experience a new partnership or revelatory encounters with a significant other this year. You could also benefit from seeing a counselor, witch, or spiritual teacher who can show you shiny new tools to add to your sharp mind.

Be ready for changes in work life March, July, and November. The changes might be painful at first, but they'll foster important growth as you learn how to value your skills and resources more. The gift of this year: to know yourself in ways you never imagined.

Rhea Wolf © Mother Tongue Ink 2018

ち ち ち xīng qī liù

♒
♓

Saturday
25

☽ ⚹ ♃ 5:51 am v/c
☽→♓ 11:07 pm

Bioregional Meditation
© Tessa Helweg-Larsen 2009

⊙⊙⊙ lǐ bài rì

♓

Sunday
26

☽ApG 6:28 am
☽ ⚹ ♅ 8:01 am
⊙□☽ 9:33 am
⊙ ⚹ ☋ 9:37 am
☽△♂ 1:06 pm

Waning Half Moon in ♓ Pisces 9:33 am PDT

May / June

Amahin—Flower Month / Athesh—Herb Month

© Paula Franco 2017

───── ☽☽☽ Henesháal—East Day ─────

♓

Monday
27

☽□☿	1:07 am
☽⚹♀	4:51 am
☽☌♆	12:39 pm
☽⚹♄	3:21 pm
☽□♃	5:54 pm
☽⚹♇	9:21 pm v/c

Salute to the Sun

───── ♂♂♂ Honesháal—West Day ─────

♓
♈

Tuesday
28

☽→♈	11:32 am

───── ☿☿☿ Hunesháal—North Day ─────

♈

Wednesday
29

☉⚹☽	2:51 am
☽□♂	4:30 am
☿□♆	6:22 pm

───── ♃♃♃ Hanesháal—South Day ─────

♈
♉

Thursday
30

☽⚹♅	1:02 am	☿⚻♄	8:19 am
☽□♄	2:16 am	♀⚹♆	9:50 am
☽△♃	4:21 am	☿☍♃	8:11 pm
☽□♇	8:08 am v/c	☽→♉	9:43 pm

───── ♀♀♀ Rayilesháal—Above Day ─────

♉

Friday
31

☽☌♅	6:26 am
♀△♄	8:26 am
☽⚹♂	4:49 pm
☿⚻♇	8:14 pm

ALL ASPECTS IN PACIFIC DAYLIGHT TIME; ADD 3 HOURS FOR EDT; ADD 7 HOURS FOR GMT

Pagan Chakra Meditation

I root down Red.
My spine, a sturdy tree,
clutches darkness, holding earth
so I may rise.
My breath, weather for my tree.
I rain, I whisper,
Orange, a flower blooms;
at uterus level, I open.
The rising sun warms,
Yellow on my trunk.
My heart branches out,
reaching Green,
my leaves touch the leaves
of those I love:
we are lissome in the wind,
a fine, clear noon.
Blue sky lifts,
with light and air,
Here I stay,
(my favorite place),
and watch for guiding birds.

At dusk
the crescent moon appears—
my third eye, Indigo,
the place between,
both blue and black,
here and there,
where answers come.
The sun now set,
Violet night reveals the stars.
They allow me passage out,
and out,
and out I go—
I follow them—
until they, too, darken.
And there, silent, lightless,
I touch the places
that were stars
and find my roots,
where I am home—
Red,
at the base of my spine.

© Hynden Walch 2017

♄♄ Yilesháal—Below Day

♉

Saturday
I

June

♀⊼♃ 2:20 am
☽⚹♆ 8:15 am
☽△♄ 10:12 am
☽☌♀ 12:55 pm
☽△♇ 3:53 pm v/c

☉☉ Hathamesháal—Center Day

♉
♊

Sunday
2

☽→♊ 4:48 am
♀△♇ 8:42 pm

A Way Through

The philodendron uncoils, boils past space and time and air
I find her stuck in her dance at a horizontal beam there
ferocious in her push, tattooing her way along the wall
sending out feelers like scouts to search a way through
undaunted vine, giant caterpillar battling a corner for supremacy
blind green hands shoving with the force of a mountain range
silently crushing all resistance from her dreaming heart.
Tomorrow and tomorrow, only this.

Hustle of her push, unfurling cusp, rocket thrust
crisis-concurring bionic sword, shoving all obstacles
soaring inside insistent green wattage
dispatched to the job of life with a moxie beyond human
Superheros pale beside this level of authority
For years she's been a mute gangbuster humming Life! Life!
Sighing softly down the length of her serpent rising body
Jungle roar of her will rages through my sleep
to bother my dreams just upstairs as she muscles her way through
Unsleeping rebel hammer set in motion, forever
And I, breathless witness of a task unfolding in sacred sovereignty
whether I watch or not, her kinetic intelligence
says to the men in suits: This is industry.
To the street-fighting activists: This is subversion!
Whose unsung might lifts my roof with a single green finger
and I'm thinking about earth protectors,
about witches created to speak for the land
And the wall that we shove against is ignorance and governance,
arrogance and war: the corporate industrial waste society.
Pray for the ceaseless green vehemence of plants
to pollinate our heart-minds and infuse us
with inspiration to keep pushing through every thwarting form,
only to insist on life that must continue.

VI. EVERYTHING UNDER THE SUN

Moon VI: June 3–July 2

New Moon in ♊ Gemini June 3; Full Moon in ♐ Sagittarius June 17; Sun in ♋ Cancer June 21

© *Robin Quinlivan 2014*

June
zhūīn

♊

Monday
3

☉☌☽ 3:02 am
☽□♆ 1:39 pm
☽☍♃ 4:34 pm

Jewel of the Night

New Moon in ♊ Gemini 3:02 am PDT

♊
♋

Tuesday
4

☽☌♉ 8:42 am v/c
☽→♋ 9:17 am
☿→♋ 1:04 pm
☽⚹♅ 5:36 pm

♋

Wednesday
5

☽☌♂ 7:48 am
☽△♆ 5:06 pm
☽☍♄ 6:28 pm

♋
♌

Thursday
6

☽☍♇ 12:00 am
☽⚹♀ 7:10 am v/c
☽→♌ 12:16 pm
☽□♅ 8:37 pm

♌

Friday
7

☿⚹♅ 7:16 am
☿□♃ 2:52 pm
☽PrG 4:20 pm
☉⚹☽ 4:47 pm
☽△♃ 9:34 pm

ALL ASPECTS IN PACIFIC DAYLIGHT TIME; ADD 3 HOURS FOR EDT; ADD 7 HOURS FOR GMT

Everyone's Meadow

Fae Magick
© Lisette Costanzo 2016

as we slept
the Spider encircles with her
spinnerets whirred. the Luna Moth
in her white velvet coat and boots
crept on Walnut bark

as we dreamt
the Barred Owl fed vole flesh
to her babes.
the Bat wobbled through
the sky under Pines. the Wren hid
in her gourd

as our spirits walked before dawn,
a line of pink appeared in the East.
White Datura closed their sweet trumpets
and Moonflowers said farewell to Sphinx Moths.
then we laid in the green meadow and rolled,
gathering up life until our skin was sticky with joy
then stood on the rim of the sky and blew it,
blew it like a billion Poppy Seeds into the hands
of every creature on Earth
speckled and wet with hope.

¤ *Stephanie A. Sellers 2017*

♄♄♄ Šanbe

♌ ♍

Saturday
8

☽□♀ 2:23 pm v/c
☽→♍ 2:45 pm
♀→♊ 6:37 pm
☽△♅ 11:16 pm

☉☉☉ Yekšanbe

♍

Sunday
9

☽✶☿ 4:33 am
☉□♆ 12:33 pm
☽✶♂ 5:45 pm
☽☍♆ 10:17 pm
☉□☽ 10:59 pm
☽△♄ 11:12 pm
☽□♃ 11:42 pm

Waxing Half Moon in ♍ Virgo 10:59 pm PDT

June
junio

Like a Giant Sunflower
I have swallowed the sun
see how I shine,
large centered, huge hearted
The sky is mine. I bask and I shine
excerpt ¤ Debra Hall 2017

ⅅⅅⅅ lunes

♍
♎

Monday
10

⊙⊼♄ 2:04 am
☽△♇ 5:01 am v/c
⊙☍♃ 8:28 am
☽→♎ 5:29 pm
☽△♀ 9:55 pm

♂♂♂ martes

♎

Tuesday
11

☽□♅ 2:03 pm
♃PrH 7:59 pm
☽□♂ 11:18 pm

☿☿☿ miércoles

♎
♏

Wednesday
12

☽□♄ 2:11 am
☽⚹♃ 2:27 am
⊙△☽ 5:57 am
☽□♇ 8:15 am v/c
☽→♏ 9:02 pm

♃♃♃ jueves

♏

Thursday
13

☽☍♅ 6:11 am
♀⚹♅ 10:10 am
⊙⊼♇ 2:45 pm
♂△♆ 11:11 pm

♀♀♀ viernes

♏

Friday
14

☽△♅ 12:33 am
☽△♆ 5:53 am
☽△♂ 6:13 am
☽⚹♄ 6:21 am
♂☍♄ 8:50 am
♂⊼♃ 8:53 am
☽⚹♇ 12:46 pm v/c

ALL ASPECTS IN PACIFIC DAYLIGHT TIME; ADD 3 HOURS FOR EDT; ADD 7 HOURS FOR GMT

Sol Sister *© Janet Newton 2017*

♏︎
♐︎

Saturday
15

☽→♐︎ 2:03 am
☽☌♀ 5:23 pm

☉☉☉ domingo

♐︎

Sunday
16

☿△♆ 4:43 am   ♃□♆ 8:22 am
☿⊼♃ 5:02 am   ☽☌♃ 12:08 pm
☿♂♄ 7:00 am   ☽□♆ 12:11 pm

MOON VI -June   103

Summer Solstice

The sun reaches its highest point of the year, like the culmination of a full moon's waxing. In order to stay steady in this full solar power, we ground ourselves by inviting the earth and the sky to meet in our bodies. As stewards, we take stock of self and world. Has an old teacher, perhaps the Dragon of Not-Enough, melted in the fires during the first half of the solar wheel? We bow and thank her before turning to discover the new teacher, who, as the waxing year gives way to the waning, will wrench our perspective wider.

Today we sit with the expansion of light, taking it in. To claim the new and larger boundary of our personal fire, we join it in ritual to that of others, and together, dance it outward. We make sacred ceremony not only for and with our immediate community, but for all our relations. The Lakota phrase *mitákuye oyásin* reminds us, "I am related to all things, and all things are related to me."

Susa Silvermarie © Mother Tongue Ink 2018

Two Drums Become One © Leah Marie Dorion 2014

Hide and Seek

Out where the rivers thaw
In the summer sun
Out where the laurel blooms
Have already begun
To spill their purple scent
All over the wild ground,
Though I'm often lost in rooms
Among trees I am always found.
We've got it wrong for so long
What's safe and what is not.
Do you remember
How to sit quiet
With a willow tree?
I will admit I had forgot.
The wild is the safest place,
Where life doesn't rest her changes
Where birth and death
Play hide and seek
And are always trading places.
This is the only kingdom
I can claim as home,
Though I enter it without a name
And I always enter it alone.
The wild is safe in a way
The tame will never know.
Come join me in my wandering
Over lichen and old bone
Lay down your fear and worry
And most of all, your phone
I promise you
We can't get lost
In the places where we belong.

Sunset at Tongue Point
© *Shelley Anne Tipton Irish 2013*

© *Emily Kedar 2014*

June
liù yuè

—— ◗◗◗ xīng qī yī ——

♐
♑

Monday
17

☉☍☽ 1:31 am v/c
☽→♑ 9:13 am
☽△♅ 7:29 pm

Full Moon in ♐ Sagittarius 1:31 am PDT

—— ♂♂♂ xīng qī èr ——

♑

Tuesday
18

♄⚹♆ 4:47 am
☿♂♂ 9:04 am
☽♂♄ 8:53 pm
☽⚹♆ 8:59 pm

—— ☿☿☿ xīng qī sān ——

♑
♒

Wednesday
19

☽☍♂ 3:22 am
☿☍♇ 3:56 am
☽♂♇ 4:17 am
☽☍☿ 4:19 am v/c
☽→♒ 7:00 pm
♂☍♇ 8:26 pm

—— ♃♃♃ xīng qī sì ——

♒

Thursday
20

☽□♅ 5:56 am

—— ♀♀♀ xīng qī wǔ ——

♒

Friday
21

Summer Solstice

☽△♀ 12:42 am
☽⚹♃ 7:02 am v/c
♆R 7:35 am
☉→♋ 8:54 am

Sun in ♋ Cancer 8:54 am PDT

ALL ASPECTS IN PACIFIC DAYLIGHT TIME; ADD 3 HOURS FOR EDT; ADD 7 HOURS FOR GMT

2019 Year at a Glance for ♋ Cancer (June 21–July 22)

Cancer, it's time to crack that outgrown shell. In January and July, the eclipses will set the stage for revelations, renovations, and renewal to your identity, which can include changes to your appearance as well as your life intentions. As you open to the cosmic potential of this year, by allowing yourself to be more connected, empathetic, and sensitive, you light the way for the rest of our world to learn the real meaning of response-ability. While those months will be busy for you, the insights exposed will take time to integrate throughout the year. To get into the rhythm of this year, remember: She changes everything she touches, and everything She touches changes.

Saturn and Pluto are helping you learn how to set clearer boundaries with a partner, lover, or close friend. This could mean some heartbreak and setbacks as you get your relationships in balance. You are incredibly sensitive to the needs of others, but sometimes go overboard in your desire to help. By confronting tendencies to overextend or meddle, you'll illuminate your own needs in relationships while encouraging others to do the same.

Jupiter brings a shift in consciousness to self-care. Consider radical departures from usual routines. Go dancing, learn horseback riding, or take a somatic awareness class. By December, your shell will reveal an iridescence born from both wearing away an old layer and building up the new you.

Rhea Wolf © Mother Tongue Ink 2018

Mera □ *Kimberly Webber 2015*

ㄣㄣㄣ xīng qī liù

♒
♓

Saturday
22

☽→♓   7:01 am
☉△☽   8:57 am
☽✶♅   6:25 pm

⊙⊙⊙ lǐ bài rì

♓

Sunday
23

☽ApG 12:46 am   ☽✶♄   8:09 pm
♀☌♃   9:45 am   ☽□♀   8:14 pm
☽□♃   7:03 pm   ☽☌♇   8:55 pm
♀⊼♄   7:32 pm

June
Athesh — Herb Month

♓
♈

Monday
24

♀□♇   2:58 am
☽⚹♇   4:17 am
☽△♂   10:22 am
☽△♅   4:10 pm v/c
☽→♈   7:38 pm

———— ♂♂♂ Honesháal — West Day ————

♈

Tuesday
25

☉□☽   2:46 am

———— ☿☿☿ Hunesháal — North Day ————   Waning Half Moon in ♈ Aries 2:46 am PDT

♈

Wednesday
26

☽△♃   6:20 am
☽□♄   7:38 am
☽⚹♀   2:38 pm
☽□♇   3:43 pm
☿→♌   5:19 pm

———— ♃♃♃ Hanesháal — South Day ————

♈
♉

Thursday
27

☽□♂   12:51 am v/c   ☉⚹♅   10:45 am
♀⚻♇   1:20 am   ☉□♎   1:00 pm
☽→♉   6:32 am   ☽♂♅   5:34 pm
☽□☿   7:23 am   ☉⚹☽   6:06 pm

———— ♀♀♀ Rayilesháal — Above Day ————

♉

Friday
28

☽△♄   4:20 pm
☽⚹♆   5:38 pm

Fiery Mary

Sun shines harsh in the dry lands
so they are turning to cinder, poor corn will never make bread;
our great blue oceans roll back and return again
carrying strange blooms and masses,
the very air is bronze, burning like weapons—
yet still the birds sing praises at dawn to Sun—
who is ember, tinder and kindle
spark of all life on this planet
the ordinary magic of light igniting our lives.

Each day her power astonishes—Fiery Mary, Young Queen of the Universe, Lady Grainne, we ask you for kindness and offer the best that we have: learning and action, intention shining with grace; our bright spirit staying strong, knowing each morning brings a new song of love.

¤ *Rose Flint 2017*

Solar Flare © *Summer Rae 2013*

ᚺᚺᚺ Yilesháal—Below Day

☿
♊

Saturday
29

☽△♇ 12:08 am
☽⚹♂ 11:38 am v/c
☽→♊ 2:09 pm
☽⚹☿ 5:51 pm

☉☉☉ Hathamesháal—Center Day

♊

Sunday
30

☽☍♃ 8:05 pm
☽□♆ 11:03 pm

Heritage

There is no scope of my lineage beyond this country
I have never met my father
And my mother dares not stretch her hands beyond this cage
Instead, bows to her oppressor every time she kneels to pray,
A peaceful captive.

Me
a mouth full of barbed wire,
A screeching chalk board,
A punch to the gut.
I am so defiant

with meditation and cowry shells
an ankh where my cross should be
I should speak more gently than this
Apologize
Ask permission,
Shouldn't wear my spine like a ship mast,
But I am split between two lands

This one that has never wanted me
And the one I can't reach
But my steps still smell like homeland
Africa etched in the air around me,
the scent strangers cling to,
Ask me where I come from,
Wanting to hear the sound of freedom—
drums on the offbeat of my sentences

Instead my tongue clinks against the chain link of my teeth

¤ *Assetou Xango 2011*

VII. WOMANFIRE

Moon VII: July 2–July 31

New Moon in ♋ Cancer July 2; Full Moon in ♑ Capricorn July 16; Sun in ♌ Leo July 22

The Motherland © *Lindsay Carron 2014*

July
zhūyīeh

♊
♋

Monday
1

☽♂♀ 2:48 pm v/c
♂→♌ 4:19 pm
☽→♋ 6:24 pm

Angel Goddess
© *Daisy Curley 2006*

———— ♂♂♂ Sešanbe ————

♋

Tuesday
2

☽⚹♅ 4:26 am
☉♂☽ 12:16 pm

Total Solar Eclipse 9:55 am PDT*
New Moon in ♋ Cancer 12:16 pm PDT

———— ☿☿☿ Cahâršanbe ————

♋
♌

Wednesday
3

☽☍♄ 12:01 am
☽△♆ 1:40 am
☽☍♇ 7:25 am v/c
♀→♋ 8:18 am
☽→♌ 8:19 pm
☽♂♂ 10:41 pm

———— ♃♃♃ Panjšanbe ————

♌

Thursday
4

☽♂♀ 2:50 am
☽□♅ 6:11 am
☽PrG 9:53 pm
☽△♃ 11:24 pm v/c

———— ♀♀♀ Jom'e ————

♌
♍

Friday
5

☽→♍ 9:25 pm

*Eclipse visible over the Americas and the Pacific

The Times, They are Changing

Forty thousand women and more are marching; their wrath is whirling through the air—hurricanes and tornadoes burst forth whenever they roar "we want our bodies back!" Earthbound allies are padding and lumbering alongside them. Together they will drum and hum, shout and stomp, whirl and dance a ritual of reclamation, magic remembered, and never seen before.

Forty thousand women and more ululate, their awful cries reach for their stars, draw down celestial allies, open their arms, legs and hearts to welcome them home. Their feet beat a timeless rhythm that quakes the earth. Their shouts, wake ancient stones and the mountains move with compassion, spark an alchemical process where earth rallies sky, reshapes our bodies whole again.

excerpt © Mari Susan Selby 2010

Celebrating Women © *Diana Rivers 1986*

᚜᚜᚜ Šanbe

♍

Saturday
6

☽⚹♀ 3:00 am
☽△♅ 7:24 am
☉⚹☽ 9:49 pm

☉☉☉ Yekšanbe

♍
♎

Sunday
7

☽□♃ 12:20 am
☽△♄ 2:04 am
☽☍♆ 4:11 am

☽△♇ 9:50 am v/c
☿R 4:14 pm
☽→♎ 11:07 pm

July
julio

The Bonfire of the Good Girl
I do not swallow down words anymore,
afraid of the fire.
See the flame in my eyes.
Feet rooted. Hips tilted.
I am a woman who burns.
excerpt © Lucy H. Pearce 2015

--- ☽☽☽ lunes ---

♎︎

Monday

♉︎

♀□♇	4:42 am	☉☌♃	8:48 am
☽✶♂	6:09 am	☽□♀	9:33 am
☽✶♉	6:37 am	♃R	2:46 pm
♀✶♅	8:32 am	☿☌♂	3:27 pm

--- ♂♂♂ martes ---

♎︎

Tuesday

9

☽✶♃	2:29 am
☉□☽	3:55 am
☽□♄	4:22 am
☉☍♄	10:07 am
☽□♇	12:35 pm v/c
♄PrH	1:30 pm

Waxing Half Moon in ♎︎ Libra 3:55 am PDT

--- ☿☿☿ miércoles ---

♎︎
♏︎

Wednesday

10

☽→♏︎	2:29 am
☽□♉	9:47 am
☽□♂	12:20 pm
☽☍♅	1:21 pm
☽△♀	6:29 pm
☉△♆	9:31 pm

--- ♃♃♃ jueves ---

♏︎

Thursday

11

♂△♃	12:29 am
☽✶♄	8:41 am
♂□♅	11:01 am
☽△♆	11:28 am
☉△☽	12:33 pm
☽✶♇	5:28 pm v/c
♇PrH	9:41 pm

--- ♀♀♀ viernes ---

♏︎
♐︎

Friday

12

☽→♐︎	8:05 am
☽△♉	2:32 pm
☽△♂	9:09 pm

ALL ASPECTS IN PACIFIC DAYLIGHT TIME; ADD 3 HOURS FOR EDT; ADD 7 HOURS FOR GMT

Fervent Fire

Stop apologizing for your anger
For your passion, for your capacity to give a damn
Vehement mothers protect their cubs from danger
While spirited activists defend their land
From sacrificing more clean water

Don't silence the prowess of your rage or pardon its animation
Let it have its celebratory triumphs
Each feeling you produce is deserving
Of your reverence

Stand tall in your fire
Let it transmute into movement
Watch it protect and serve

Listen to it while it reminds you
How alive you are
How connected you are

And when your fire simmers
Beg to understand the other side
For when we can understand each other
Through all our complexities and disagreements
We can begin to meet in the middle

Goddess
of the Seasons
© *Joan Zehnder 2015*

This is where we grow
And build together again.

◻ *McKenzie Brill 2017*

ካካካ sábado

♐

Saturday
13

☽☌♃   1:11 pm
☽□♆   6:30 pm v/c

☉☉☉ domingo

♐
♑

Sunday
14

☉☍♇   7:51 am
☽→♑   4:05 pm

July
qī yuè

♑

Monday
15

☽△♅ 4:08 am
☽☌♀ 9:42 pm

© Sudie Rakusin 1989

♂♂♂ xīng qī èr

♑

Tuesday
16

☽☌♄ 12:18 am
♀⊼♃ 12:44 am
☽⚹♆ 3:52 am
☽☌♇ 10:16 am
☉☍☽ 2:38 pm v/c
♀☍♄ 10:34 pm

Partial Lunar Eclipse 11:55 am PDT*
Full Moon in ♑ Capricorn 2:38 pm PDT

☿☿☿ xīng qī sān

♑
♒

Wednesday
17

☽→♒ 2:19 am
☽☍♉ 4:39 am
☽□♅ 2:54 pm
☽☍♂ 10:50 pm

♃♃♃ xīng qī sì

♒

Thursday
18

☿PrH 3:54 am
☽⚹♃ 8:53 am v/c
♀△♆ 11:03 am

♀♀♀ xīng qī wǔ

♒
♓

Friday
19

☿→♋ 12:06 am
☽→♓ 2:19 pm

---

* Eclipse Visible over Europe, Asia, Aus., Africa, the Americas, Pac., Atl.& Ind. Oceans, Antarctica

I Wear My Skirts Long

I wear my skirts long, long to the ground, to the earth I love— heavy cotton, full sweep velvet, or silk, light in the wind. I walk in all weathers.. I wander in the rain and relish the weight of these skirts, the gentle pull of fabric over my hips as the wet earth soaks the hems at my feet. I like the dragging, the swish . . . the accumulating of the rogue brambles and twigs—as if, walking long enough over moor, through wood, along shore and lane, I might become full covered and camouflaged by these stray gifts of the land.

I wear nothing underneath, not a stitch but my own treasure. Lift the skirt to calm the storms, the earthquakes, the tsunamis, the vast scales of madness—one force of nature to another. Under these long skirts I am free, gentle air between my thighs. Air kissed and earth blessed, water born and fire forged. So we all are. Heart beat—yoni beat—earth beat all together as I stand tall above the ground, as I sit gratefully in the lap of this wild earth, mama of us all.

I can feel all the women before that walked this earth dressed this way, long skirts over potent wisdom, lifetime after lifetime. I hear their voices in the swishing rhythm of my skirts—*Daughter,* they say, *feel the love singing in your veins, the hands supporting at your back. Feel the pain and brilliance of life. Hear the song that calls you on. Daughter,* they say, *oh daughter: find the rhythms in your blood and bones. Daughter,* they say—insistent, urgent—*take it, take this gift. Life giving life giving life through so many lifetimes and you, you are the survival, the thriving of all that came before. All we know is yours. Take it, daughter, receive our blessing.*

© *Nell Aurelia 2015*

--------------- ㅅㅅㅅ xīng qī liù ---------------

☿ ◗ **Saturday**
20

☽⚹♅ 3:16 am
☽ApG 5:00 pm
☽□♃ 9:06 pm
☽⚹♄ 11:28 pm

--------------- ☉○○ lǐ bài rì ---------------

☿ ◗ **Sunday**
21

♀☍♇ 1:32 am
☽☌♆ 3:45 am
☉☌☿ 5:34 am

☽⚹♇ 10:18 am
☽△♀ 11:20 am
☽△☿ 10:58 pm

July
Ameda—Vegetable Month

♓
♈

Monday
22

☉△☽ 1:34 am v/c
☽→♈ 3:02 am
☉→♌ 7:50 pm

━━━━━ ♂♂♂ Honesháal—West Day ━━━━━ Sun in ♌ Leo 7:50 pm PDT

♈

Tuesday
23

☽△♂ 6:34 am
☽△♃ 9:14 am
☽□♄ 11:30 am
☽□♇ 10:20 pm

━━━━━ ☿☿☿ Hunesháal—North Day ━━━━━━━━━━━

♈
♉

Wednesday
24

☽□♀ 6:12 am
☽□♅ 7:48 am v/c
☽→♉ 2:42 pm
☿♂♀ 5:26 pm
☉□☽ 6:18 pm

━━━━━ ♃♃♃ Hanesháal—South Day ━━ Waning Half Moon in ♉ Taurus 6:18 pm PDT

♉

Thursday
25

☽♂♅ 3:16 am
♂△♃ 5:22 am
☽□♂ 8:11 pm
☽△♄ 9:29 pm

━━━━━ ♀♀♀ Rayilesháal—Above Day ━━━━━━━━━━

♉
♊

Friday
26

☽⚹♆ 1:59 am
☽△♇ 7:57 am
☽⚹♅ 2:43 pm
♂⚻♄ 7:41 pm
☽⚹♀ 9:28 pm v/c
☽→♊ 11:29 pm

ALL ASPECTS IN PACIFIC DAYLIGHT TIME; ADD 3 HOURS FOR EDT; ADD 7 HOURS FOR GMT

2019 Year at a Glance for ♌ Leo (July 22–Aug 23)

2019 starts out with a bang for you, with a lunar eclipse in Leo—the last one for 19 years. Yet much of the year will focus on self-care and surrender. When we're looking up at the Sun's bright golden light, it can be easy to forget that it's one star among billions. This year, you're invited to be "one among many" rather than "the chosen one." You are learning to build a better inner world, which includes tending to your health and your spiritual practice.

Uranus will initiate you into ways of working that are more sustainable and collaborative, but you may first experience some burn-out. Go behind-the-scenes and get organized. Allowing others to work in the front lines as you rejuvenate is an important activist skill. The key here is service rather than applause. Pluto and Saturn are teaching you how serving others, even when you don't get any credit, brings wisdom and ultimately, allows you more space to take care of your physical and emotional needs.

Although your work may be less visible this year, Jupiter brings a shift in consciousness to the topics of creativity, children, and romance. Projects you've struggled with in the past will take off. You'll be motivated to express yourself in original ways, as an artist, parent, or lover. The energizing nature of this transit may lead you to take on too much or take unnecessary risks. Remember to protect your body and get enough rest.

Rhea Wolf © Mother Tongue Ink 2018

She Brings the Sun
© Chasity Bleu 2017

♄♄♄ Yileshą́ą́l—Below Day

♊ Saturday 27

☉⚹☽ 7:16 am
♀→♌ 6:54 pm

☉☉☉ Hathameshą́ą́l—Center Day

♊ Sunday 28

☽☍♃ 2:08 am
☽⚹♂ 5:45 am
☽□♆ 8:24 am v/c
☉△⚷ 8:43 pm

Incredible Edible Todmorden

"The industrial revolution came . . . and went."
Thus begins the story of Todmorden, England,
the little town that could.
Food grows free for the picking, everywhere,
at the police station, the fire house, the schools.
Yum, yum! Fresh and free, festivals and street fairs,
recipes traded from around the world.
All grown here or right nearby.
"Everyone's got to eat," they say, and so they do!
"The time to act is now."
Creating a world truly nourishing,
for their children, for us all.
Food production begins in the garden of every school,
vegetables, chickens and fruit trees.
"The joy of connecting people is fabulous."
Training the young people to grow food and market it,
small sustainable jobs where despair and depression had been.
In every nook and cranny, an apple tree.
"Go ahead, take some, it's free!"
Poultry raising, bee keeping, dairy.
"You just have to give a damn about tomorrow."
Dear little Todmorden, voting for life with all your being,
juggling community, education, and business—
keep those three plates spinning in the air!

Seed Mother © *Penn 2017*

July / August
zhūyīeh / ut

—— ⟩⟩⟩ Došanbe ——

♊
♋

Monday
29

☽→♋ 4:31 am
☽✶♅ 3:36 pm
☉□♅ 4:14 pm

—— ♂♂♂ Sešanbe ——

♋

Tuesday
30

☽☌♄ 6:57 am
☽△♆ 11:14 am
♂⚹♆ 1:09 pm
☽☍♇ 4:27 pm
☽☌☿ 8:32 pm v/c

—— ☿☿☿ Cahâršanbe ——

♋
♌

Wednesday
31

☽→♌ 6:18 am
☽☌♀ 1:51 pm
☽□♅ 4:54 pm
☉☌☽ 8:12 pm
♅D 8:58 pm

New Moon in ♌ Leo 8:12 pm PDT

—— ♃♃♃ Panjšanbe ——

♌

Thursday
1

August

☽△♃ 5:53 am
♀△♇ 9:58 am
☽☌♂ 1:48 pm v/c

—— ♀♀♀ Jom'e ——

♌
♍

Friday
2

Lammas / Lughnasad

☽PrG 12:05 am
♀□♅ 3:00 am
☽→♍ 6:20 am
☽△♅ 4:50 pm

ALL ASPECTS IN PACIFIC DAYLIGHT TIME; ADD 3 HOURS FOR EDT; ADD 7 HOURS FOR GMT

Lammas

We wombstorm all the creative ways we might celebrate first harvest. One is belly dancing. Half of our dopamine and most of our serotonin, hormones associated with good feelings, are produced in the enteric brain in our bellies. Since the womb is adjacent to the small intestine, that gut feeling is more likely our delphic intuition.

Everything we wish to create in our world gestates inside of us. Our essence has ripened into a larger version of itself during the year's turnings. Now we harvest both the fruits of Her womb and the direct knowing of our own.

The natural world is thriving with excitement and magic, and yet we feel aware that everything will die soon. Hail to Tonantzin, Ceres, Pomona, Demeter. We reap, feast, make corn dolls and work magic in the world. Let's cast a spell against corporate moves to grow GMO corn in Mexico, the cradle of 59 varieties of native corn. We give thanks, and above all, we give back to our planet.

Susa Silvermarie © Mother Tongue Ink 2018

Garden Project *© Jakki Moore 2017*

Old Singing Woman © Rita Loyd 1999

Original Sun Power

Bidden, Beckoned, Called
Conjured, Cast, Played, Willed
Summoned by Women!

Ritual, Ceremony, Singing
Drumming, Shaking, Dancing

Coal Stirring, Soul Purring, Incantation Sparking of Women!

All hands entwine around a great table of Women's Love
A Meal prepared from Sun-Fire
Our Tongues burning with Truth
Un-Extinguishable, Eternal
Original Source of All That Is
Burning in the Heart of our Flaming Souls
Nourishing Embers of Wisdom
From the Radiant—Hot Creations
Of Women!

¤ *Stephanie A. Sellers 2017*

Enchanted Table

See this old banquet table
Made of sacred oak wood
From a small Mediterranean island
Where the people still worship Artemis.
Legend has it that Saint Joan once ate here,
Gathering strength
Before she went to capture Rheims.

A well-used table.
Old scars worn smooth by time
And women's hands,
With new nicks and scratches that
Testify to present use.

The dark-grained satin wood
Beckons my cheek
To feel the energy of past meals
Flow and gather within me.

Goddess,
Give me strength
For I too
Must capture Rheims.

□ *Susan L. Roberts 1980*

—— ꘈꘈꘈ Šanbe ——

♍ **Saturday**
3

☽□♃ 5:39 am
☽△♄ 7:03 am
☽☍♆ 11:30 am
☽△♇ 4:35 pm
☽⚹♅ 9:27 pm v/c

—— ☉☉☉ Yekšanbe ——

♍
♎ **Sunday**
4

☽→♎ 6:30 am
♂⊼♇ 8:59 am
☽⚹♀ 11:00 pm

August
agosto

Monday
5

☉✶☽	3:25 am
☽✶♃	6:27 am
☽□♄	7:46 am
☽□♇	5:51 pm
☽✶♂	7:26 pm

On Support
Support feels like
A warm bath on tired skin
Long arms hugging in.
excerpt © Robin D. Bruce 2017

Tuesday
6

☽□☿	12:36 am v/c
☽→♏	8:31 am
☽☍♅	7:55 pm

Wednesday
7

☉△♃	12:31 am
☽□♀	7:01 am
☉□☽	10:31 am
☽✶♄	11:02 am
☽△♆	4:15 pm
☉⊼♄	5:24 pm
☽✶♇	9:53 pm

Waxing Half Moon in ♏ Scorpio 10:31 am PDT

Thursday
8

☽□♂	2:16 am
☽△☿	7:58 am v/c
♀△♃	1:27 pm
☽→♐	1:35 pm

Friday
9

♀⊼♄	1:23 am
☽♂♃	4:25 pm
☽△♀	7:18 pm
☉△☽	9:39 pm
☽□♆	11:11 pm

Allies in the Storm © Denise Kester 2015

♐
♑

Saturday
10

☽△♂ 12:50 pm v/c
☉⚹♆ 5:44 pm
♀ApH 7:46 pm
☽→♑ 9:50 pm

♑

Sunday
11

♃D 6:37 am
☽△♅ 10:37 am
♀⚹♆ 10:43 am
☿→♌ 12:46 pm
♅R 7:27 pm

August
bā yuè

Brighid's Altar
◻ Beth Lenco 2017

───── ☽☽☽ xīng qī yī ─────

♑ ☽ ## Monday
12

☽☌♄ 2:53 am
☽⚹♆ 8:58 am
☽☌♇ 3:11 pm v/c

───── ♂♂♂ xīng qī èr ─────

♑ ☽ ## Tuesday
♒ ## 13

☽→♒ 8:35 am
☽☍♅ 1:33 pm
☽□♅ 9:47 pm
☉☌♀ 11:07 pm
♀⊼♇ 11:08 pm
☉⊼♇ 11:08 pm

───── ☿☿☿ xīng qī sān ─────

♒ ☽ ## Wednesday
14

☽⚹♃ 1:37 pm

───── ♃♃♃ xīng qī sì ─────

♒ ☽ ## Thursday
♓ ## 15   Lunar Lammas

☉☍☽ 5:29 am
☽☍♀ 6:16 am
☿△♄ 2:15 pm
☽☍♂ 6:02 pm v/c
☽→♓ 8:49 pm

───── ♀♀♀ xīng qī wǔ ─────

Full Moon in ♒ Aquarius 5:29 am PDT

♓ ☽ ## Friday
16

☿□♅ 10:07 am
☽⚹♅ 10:11 am

ALL ASPECTS IN PACIFIC DAYLIGHT TIME; ADD 3 HOURS FOR EDT; ADD 7 HOURS FOR GMT

Home Fires Burn

I am a fire keeper, a woman of the flame. I burn and I blaze; I incinerate, and I light up. My bones are infused with sparks; my marrow knows flint and friction. My arms carry woods from the forest: oak, manzanita, madrone, and redwoods that burn brilliant at my fireplace, where my kettle boils and volatile aromas delight noses. My fingers prepare tinder bundles from cattail duff, mugwort leaves, pine needles, and tree's blood. My eyes shine with the light of thousands of suns, rising and setting as I bring fire into my body. I am joined by my fiery kin each time I tend my sacred flame, for my sisters and ancestors know the ways of the heat. Our cauldrons have bubbled with healing potions and nourishing traditions, our smudges have cleared the air and prepared us for ritual, and our campfires have drawn us together to warm our souls with community. By the light of the flickering reds and alive oranges I feel myself ignited, my creative passions burning within to match my Hestia-honored hearth.

¤ *Mori Natura 2017*

━━━ ♄♄♄ xīng qī liù ━━━

♓

Saturday
17

☽□♃ 2:17 am
☽✶♄ 2:35 am
☽ApG 3:49 am

☽♂♅ 9:08 am
☽✶♇ 3:34 pm v/c
♂→♍ 10:18 pm

━━━ ☉☉☉ lǐ bài rì ━━━

♓
♈

Sunday
18

☽→♈ 9:33 am

August
Adaletham—Berry Month

─── ☽☽☽ Henesháal—East Day ───

♈

Monday
19

☽△♅ 8:41 am
☽□♄ 2:56 pm
☽△♃ 2:58 pm

─── ♂♂♂ Honesháal—West Day ───

♈
♉

Tuesday
20

☽□♇ 3:53 am
☉△☽ 5:01 pm
☽△♀ 9:06 pm v/c
☽→♉ 9:37 pm

─── ☿☿☿ Hunesháal—North Day ───

♉

Wednesday
21

☿⊼♄ 1:26 am
☽△♂ 1:33 am
♀→♍ 2:06 am
☿△♃ 3:05 am
☽☌♅ 10:33 am

─── ♃♃♃ Hanesháal—South Day ───

♉

Thursday
22

☽△♄ 1:57 am
☽□♉ 6:22 am
☽✶♆ 8:23 am
☽△♇ 2:32 pm v/c
☿⊼♆ 7:26 pm

─── ♀♀♀ Rayilesháal—Above Day ───

♉
♊

Friday
23

☉→♍ 3:02 am
☽→♊ 7:34 am
☉□☽ 7:56 am
☽□♀ 1:17 pm
☽□♂ 2:18 pm

Sun in ♍ Virgo 3:02 am PDT
Waning Half Moon in ♊ Gemini 7:56 am PDT

─────────────────────────────
ALL ASPECTS IN PACIFIC DAYLIGHT TIME; ADD 3 HOURS FOR EDT; ADD 7 HOURS FOR GMT

2019 Year at a Glance for ♍ Virgo (August 23–Sept. 23)

2019 encourages you to step out to engage the community and share your passion. The spring brings fiery flashes of synchronicities, propelling your spiritual life out of its rut and opening you to exciting journeys both literal and metaphoric. Pay attention to the signs, and you will find wild meaning and intention for your life.

Although communications with your significant others may feel strained or confusing in March, resist the urge to over-explain things. This year, Mercury is illuminating the difficulties and talents of your social self—how you communicate, negotiate, and collaborate with others. You don't have to let everyone know your feelings, but it's imperative that *you* know how you're feeling. Take time to do an emotional check-in before meetings, gatherings, or dates.

Jupiter brings the consciousness shift close to home. You may have extra energy or resources to revamp your dwelling space. Let your creativity shine as you build a deeper connection to where you live and who you live with.

Speaking of creativity, Saturn and Pluto have been pressuring you to make something, express yourself, or fight for your right to have fun. This energy can be compulsive or destructive when ignored. Instead, cultivate health and integrity by making time and space for self-expression. It doesn't matter how much you create or who sees it. Don't repress your playfulness; it is revolutionary and transformative, and will add zest to any community action groups you are involved in.   *Rhea Wolf © Mother Tongue Ink 2018*

Two of Fire
© Jan Kinney 1999

──────── ♄♄♄ Yileshàal—Below Day ────────

   ♊

Saturday
24

♀☌♂ 10:04 am
☽☍♃ 10:53 am
☿⊼♇ 11:14 am
☽□Ψ   4:17 pm
☽⚹☿ 11:58 pm v/c

──────── ☉☉☉ Hathameshàal—Center Day ────────

   ♊
♋

Sunday
25

♀⊼♄ 3:33 am
☽→♋ 2:05 pm
☉⚹☽ 6:33 pm
♂⊼♄ 7:12 pm
☽⚹♂ 10:59 pm

August / September
ut / september

© Sophia Rosenberg 2008

Brigid's Blessing

))) Došanbe

♋

Monday
26

)✶♀ 12:44 am
)✶♅ 1:27 am
♀△♅ 8:38 am
)☍♄ 2:45 pm
)△♆ 8:28 pm

♂♂♂ Sešanbe

♋
♌

Tuesday
27

)☍♇ 1:55 am v/c
)→♌ 4:53 pm

☿☿☿ Cahâršanbe

♌

Wednesday
28

)□♅ 3:29 am
♂△♅ 3:53 am
☉⚻♏ 5:17 am
♂ApH 12:23 pm
)△♃ 5:07 pm v/c

♃♃♃ Panjšanbe

♌
♍

Thursday
29

☿→♍ 12:48 am
)→♍ 4:57 pm
)☌☿ 7:22 pm
☉△♅ 8:14 pm

♀♀♀ Jom'e

♍

Friday
30

)△♅ 3:09 am   )☌♀ 11:13 am
☉☌) 3:37 am   )△♄ 3:14 pm
)☌♂ 5:15 am   )□♃ 4:39 pm
)PrG 8:59 am   )☍♆ 8:35 pm

New Moon in ♍ Virgo 3:37 am PDT

All aspects in Pacific Daylight Time; add 3 hours for EDT; add 7 hours for GMT

Migration Creation © *Toni Truesdale 2017*

ᚼᚼᚼ Šanbe

♍
♎

Saturday
31

☽△♇ 1:46 am v/c
☿⊼♂ 10:57 am
☽→♎ 4:08 pm

☉☉☉ Yekšanbe

♎

Sunday
1

September

☿△♅ 7:11 am
♀△♄ 11:49 am
☽□♄ 2:40 pm
☽⚹♃ 4:24 pm

Water Protectors

The Missouri pours the longest river in America,
flows into the Mississippi near the birthplace of *Black Lives Matter*.
It flows past where Katrina washed a city clean of poor people
flows into the Atlantic Ocean where five centuries ago,
slave-stealing ships followed smoke of burning witches,
paving the road for Christianity/science/capitalism
Colonial schism erasing pagan cultures.
Victors forget how they get their spoils, but the de-spoiled remember
Histories flow together like rivers and pipelines. Water remembers.

Navajo social media fairy said prayer is simple: *Help me.*
Thank you. And off she blew to The Protectors' Dakota showdown.
Before there was Facebook, Deep-Time people sent out drum stories,
ancestors of these same horses whinnied, prancing under an older sky
when the invaders shot them all down. The river flows on.

Home, writing poems in a room full of snakes—
Rushing whitewater beams from Facebook Hill
where they're indigenizing media with digital smoke signals.
Live feeds story-tell prophesies; black snake devouring the world.
Capitalism's rivers choking on pipelines—*The whole world is watching*

It's simple, she said: *Help me.* Histories flow together. *Thank you.*
Elders sit in chairs, while warriors gallop the endless body of wind
flaming arrows sending pixels across their land
the raised fists fires feathers drums, the tanks teargas babies army
the calm, beautiful horses. *Mni Wiconi.* Water is life.
Ceremony broadcast over a bandwidth of wind and eagles
My bones the rattle, my heart the drum
I'm pierced with visions of earth and water, fire and air
Great Mystery of living rivers-forests-animals. A fierce rudder, a blade.
My voice and arms raised with theirs.
Prayer, a net tossed over all there is.

IX. CARRYING THE TORCH

Moon IX: August 30–September 28

New Moon in ♍ Virgo Aug. 30; Full Moon in ♓ Pisces Sept. 13; Sun in ♎ Libra Sept. 23

The Pact
© *Autumn Skye ART 2016*

September
septiembre

☽☽☽ lunes

♎ ♏

Monday
2

☽□♇ 1:34 am v/c
☉☌♂ 3:42 am
♀□♃ 9:26 am
☽→♏ 4:35 pm

MuthaQueen Cypher

¤ *IAYAALIS Kali-Ma'at Eloai 1996*

♂♂♂ martes

♏

Tuesday
3

☽☍♅ 3:18 am
☿☌♂ 8:40 am
☽⚹♂ 10:12 am
☽⚹☿ 10:21 am
☉⚹☽ 10:58 am
☽⚹♄ 4:18 pm
☉☌☿ 6:40 pm
☽⚹♀ 9:31 pm
☽△♆ 10:09 pm

☿☿☿ miércoles

♏ ♐

Wednesday
4

☽⚹♇ 3:58 am v/c
♀☍♆ 4:26 am
☽→♐ 8:08 pm

♃♃♃ jueves

♐

Thursday
5

☿△♄ 5:37 am
☽□♂ 5:49 pm
☉□☽ 8:10 pm

Waxing Half Moon in ♐ Sagittarius 8:10 pm PDT

♀♀♀ viernes

♐

Friday
6

☿□♃ 12:11 am
☽☌♃ 12:20 am
☽□☿ 12:21 am
☽□♆ 3:52 am
☽□♀ 9:03 am v/c
☉△♄ 2:56 pm
♀△♇ 8:46 pm

ALL ASPECTS IN PACIFIC DAYLIGHT TIME; ADD 3 HOURS FOR EDT; ADD 7 HOURS FOR GMT

Outrage

. . . well you aught to be outraged
you oughtta be *outrageously* outraged.

For outrage is the fierce wild unequivocal love
that burns in a wolf-mother's eyes.
It is the feral opposite of shame
devouring deception in its crackling flames.
It is the stone that sharpens swords to pierce illusion,
it's the force that parts the morning's curtains
flooding the world with fresh horizons.

Outrage is a potent spell to summon belief in the face of disgrace:
It is a howl of empathy with all the lives whose tongues are tied—

Sweetheart: dig through the ruins of our shared yesterdays:
Witch burnings. Blitzkriegs. Gas chambers. Firebombs.
There are cinders among these ashes:
Stir up your remembering. Blow life into that ruckus.

Let outrage unleash your cantankerous truth
Let it spill off your tongue like sacramental wine.
Allow it to propel you out of your chair, and into the world—
May sparks of outrage guide you out of cold indifference
towards compassion's crimson glow.

Sound your outrageous yawp, like a trumpet peal
to rouse honour from its sleep and slap devotion into roaring life.

¤ *Andrea Palframan 2017*

ħħħ sábado

♐
♑
Saturday
7

☿☍♆ 12:18 am
☽→♑ 3:37 am
☽△♅ 3:47 pm

☉☉☉ domingo

♑
Sunday
8

☽△♂ 5:53 am   ☽△♅ 8:02 pm
☽♂♄ 6:42 am   ☽♂♇ 8:03 pm
☉□♃ 8:26 am   ☿△♇ 8:09 pm
☉△☽ 10:11 am   ♂△♄ 9:14 pm
☽✶♆ 1:18 pm

September
jiǔ yuè

Full Circle

———— ☽☽☽ xīng qī yī ————

♑
♒

Monday
9

☽△♀ 1:30 am v/c
ΨPrH 4:02 am
☽→♒ 2:24 pm

———— ♂♂♂ xīng qī èr ————

♒

Tuesday
10

☉☌Ψ 12:24 am
☽□♅ 2:57 am
☿ApH 11:40 am
☽✶♃ 10:22 pm v/c

———— ☿☿☿ xīng qī sān ————

♒

Wednesday
11

No Exact Aspects

———— ♃♃♃ xīng qī sì ————

♒
♓

Thursday
12

♂□♃ 2:06 am
☽→♓ 2:51 am
☽✶♅ 3:27 pm

———— ♀♀♀ xīng qī wǔ ————

♓

Friday
13

☽ApG 6:32 am   ☽☌♂ 1:12 pm
☽✶♄ 7:04 am   ☽☌Ψ 1:43 pm
☿☌♀ 8:11 am   ☽✶♇ 8:49 pm
☽□♃ 11:35 am   ☉☍☽ 9:33 pm v/c
☉△♇ 12:42 pm   ♂☍Ψ 10:25 pm

Full Moon in ♓ Pisces 9:33 pm PDT

ALL ASPECTS IN PACIFIC DAYLIGHT TIME; ADD 3 HOURS FOR EDT; ADD 7 HOURS FOR GMT

Justice & Mercy © Sue Ellen Parkinson 2017

─── ꜜꜜꜜ xīng qī liù ───

♓
♈

Saturday
14

☿→♎ 12:14 am
♀→♎ 6:43 am
☽→♈ 3:32 pm
☽☍♀ 4:34 pm
☽☍☿ 6:08 pm

─── ⊙⊙⊙ lǐ bài rì ───

♈

Sunday
15

☽□♄ 7:29 pm

September

Ahede—Grain Month

----- ☽☽☽ Henesháal—East Day -----

♈ Monday
16

☽△♃ 12:29 am
☽□♇ 9:03 am v/c
☿☍♄ 9:55 am

----- ♂♂♂ Honesháal—West Day -----

♈
♉ Tuesday
17

☽→♉ 3:31 am
♀☍♄ 1:21 pm
☿⊼♅ 1:56 pm
☽♂♅ 3:31 pm

----- ☿☿☿ Hunesháal—North Day -----

♉ Wednesday
18

♄D 1:47 am
☽△♄ 6:52 am
☽⚹♆ 1:02 pm
☽△♂ 7:19 pm
☽△♇ 8:02 pm

----- ♃♃♃ Hanesháal—South Day -----

♉
♊ Thursday
19

♀⊼♅ 3:27 am
☉△☽ 6:57 am v/c
♂△♇ 8:53 am
☽→♊ 1:58 pm

----- ♀♀♀ Rayilesháal—Above Day -----

♊ Friday
20

☽△♀ 3:50 am
☽△♅ 10:24 am
☽☍♃ 9:54 pm
☽□♆ 10:02 pm

ALL ASPECTS IN PACIFIC DAYLIGHT TIME; ADD 3 HOURS FOR EDT; ADD 7 HOURS FOR GMT

About My Name . . .

Many of my role models
Contemporaries
But especially,
Ancestors
Have a name that brings the tongue to worship.
Names that feel like ritual in your mouth.

I am finished with names that leave me unmoved
Names that leave the speaker's mouth unscathed
I want a name like fire
Like rebellion
Like my hand gripping massa's whip

I want a name from before the ships
I want a name that catches you in the throat when you say it wrong
And if you're afraid to say it wrong
Well then I guess you should be

I want a name that only the brave can say
One that only sits right in the mouth of those who love me right
Because only the brave can love me right

◻ Assetou Xango 2016

──────── ˥˥˥ Yilesháal—Below Day ────────

♊
♋

Saturday
21

☽□♂︎   7:05 am
♃□♆   9:44 am
☉□☽   7:41 pm v/c
☽→♋   9:50 pm

Waning Half Moon in ♊ Gemini 7:41 pm PDT

──────── ⊙⊙⊙ Hathamesháal—Center Day ────────

♋

Sunday
22

☽✳︎♅   8:31 am
♉□♄   9:19 am
☽□♀   4:30 pm
☽♂♄ 10:38 pm

Autumn Equinox

The Mother Sun shines directly on the Equator again, and dances our day and night into balance. Since Summer Solstice, the hours of daylight have been greater than the hours from dusk to dawn. For the rest of the Wheel of the Year, the reverse holds true.

In this final harvest, we have finished old projects and dreams. Trees releasing their leaves show us how to let go. Before turning from summer and facing winter, we reflect on what we are reaping, especially as a human family. We actively seek and uphold that which is positive.

The rhythm of the solar wheel suggests rest. We slow down to gently observe, absorb, and rejoice in Gaia. During the transition called autumn when we celebrate both bounty and impermanence, we reflect that Gaia too, is impermanent. How much time does She have? Yet She asks for trust, even given the alarms that surround us. We and the world flare in brief brilliance—and embrace the first footsteps of darkness.

Susa Silvermarie © Mother Tongue Ink 2018

Foxy Sunyata, Rainbow in the Void © Lindy Kehoe 2017

Red River Rendezvous
© Leah Marie Dorion 2014

The Village Life

There is a road. It winds. The way is everywhere. Where you place your feet, tiny villages of lichen and bone are born. Where you breath as you walk, birds sprout on the ankles of clouds. The gods are watching. Feed them your life.

Thunder claps visit each afternoon. Between chores, we pause to listen. We fill our own bellies with water, we would drink the Lightning if we knew how. We would weave each drop of water to make a boat that cannot sink, will not sink as long as form persists. We would grow each seed in our cheeks, trailing thick green vines from our open mouths in summer.

I lay in reverie, bound within a body in a broken time. This is one story that we tell. But even as we savor it, the story ends, broken time sits on its hind legs and looks skyward, having licked all its wounds to gold. We go off in all directions without ever leaving home. The sun's light pours into our skin, gives birth to our vegetable and animal kin. We know that it's inside this fire that all life begins.

Dinner bells resound through the mountain on a day between the seasons. The rain of early winter is here though the peaches on the tree are green. The sun is missing. We take what comes. The season's rhythms begin to syncopate, change without warning. Perhaps the wise among us have some new story to keep us tethered to the cycles. The village matters now, more than ever. The world turns on a thread spun from grief and love, a thread dipped in blood and dried under the breath of song. The world will spin long after humans have outlived their usefulness. Rise up now. Remember.

© Emily Kedar 2015

September
septämber

── ꝺꝺꝺ Došanbe ──────────────

Monday
23

Fall Equinox

☽□♅ 12:23 am
☉→♎ 12:50 am
☽△♆ 3:55 am
☽⚼♇ 10:21 am
☽⚹♂ 3:05 pm v/c

Sun in ♎ Libra 12:50 am PDT

── ♂♂♂ Sešanbe ──────────────

Tuesday
24

☽→♌ 2:19 am
☉⚹☽ 4:12 am
☿⚻♆ 6:57 am
☽□♅ 12:12 pm
☿⚹♃ 2:00 pm

── ☿☿☿ Cahâršanbe ──────────────

Wednesday
25

☽⚹♀ 12:35 am
♄PrH 4:54 am
☽△♃ 7:21 am
☽⚹☿ 9:14 am v/c
♀□♄ 12:20 pm

── ♃♃♃ Panjšanbe ──────────────

Thursday
26

☽→♍ 3:37 am
☽△♅ 12:53 pm
☿□♇ 4:52 pm
☉⚼♄ 6:21 pm

── ♀♀♀ Jom'e ──────────────

Friday
27

☽△♄ 1:49 am
☽⚼♆ 6:21 am
☽□♃ 7:45 am
☽△♇ 12:20 pm
☽PrG 7:11 pm
♀⚻♆ 7:50 pm
☽♂♂ 8:58 pm v/c

ALL ASPECTS IN PACIFIC DAYLIGHT TIME; ADD 3 HOURS FOR EDT; ADD 7 HOURS FOR GMT

2019 Year at a Glance for ♎ Libra (Sept. 23–Oct. 23)

A big theme of this year will be work/life balance. It's time to purge any worn-out ideas you have about home. Although a strong foundation is vital when creating shelter, your ideas of home don't have to be rigid, old-fashioned, or stunted by too many rules.

Pluto and Saturn are pressuring you to get specific about what is essential for you to feel rooted—and turn the rest into compost. The work you are called to do—whether paid or unpaid—can be a source of renewal and vitality this year. Could it be that what you do out in the world is a form of home for you? How is work both nourishing to you and supportive of others?

Uranus is burning old structures in the deepest parts of you. People spend a lot of time trying to figure out "who they are." But the gift of this year includes *un*knowing who you are. As Octavia Butler writes: "In order to rise from its own ashes, a Phoenix first must burn." Ultimately, this breakthrough of unseen potential in late spring will shine the light on healthy patterns of intimacy and inner growth.

To activate your consciousness shift in 2019, invite Jupiter to give you a new name that inspires confidence or take a class to learn an empowering skill. Get to know your neighbors and their dreams. Write a heartfelt letter to a sibling. There are fresh sources of inspiration just waiting in plain sight.

Rhea Wolf © Mother Tongue Ink 2018

Liberty
© *Autumn Skye ART 2016*

— ♄♄♄ Šanbe —

♍
♎

Saturday
28

☽→♎ 3:03 am
☉☌☽ 11:26 am
♀⚹♃ 4:40 pm
☉⚻♅ 9:14 pm

— ☉☉☉ Yekšanbe —

New Moon in ♎ Libra 11:26 am PDT

♎

Sunday
29

☽□♄ 1:08 am
☽⚹♃ 7:27 am
☽☌♀ 8:38 am
☽□♇ 11:40 am
☽☌♅ 7:06 pm v/c

On Becoming a Crone

aging is not stealthy, like some ninja warrior sneaking up and nunchucking me from behind. new lines and wrinkles appear in the mirror daily whether I choose to acknowledge them or remain in denial. I am surprised when I bend down to pick up something I've dropped again to feel a new twinge or stab catch my breath.

aging is not beautiful; the silver framing my face may be precious, but only as a reward for all the courage I've revealed. the soft curves of my body are not a sign of vulnerability, but symbolize the great expanse of my soul. today I treasure really looking people in the eye, saying aloud, "I like your smile, your sweater, your care."

aging is not inevitable; we are blessed with each bonus day, another moment to tell a loved one they are beloved. I try to halt the peevishness I feel towards the flesh hanging from my arms by assuring myself that I now have angels' wings, and then I laugh at myself because crones aren't always angels. sometimes we transform into bitches, refueling our wrath, and when that happens, I whisper, "can you see me now?"

aging is not in the mind where I will always remain 30, while my body changes and prepares for the next transition. I forgive myself for the days when I wheeze like an organ needing repair, for truly I am an oak tree with strong roots, able to dance in the wind, yet curious, wiley and wild enough to March on Washington while proudly shouldering a rainbow flag a little higher. and when that happens can you hear me roar?

silver power is not for the faint hearted; look us in the eyes if you dare, discount us at your peril. you will see tigers baring their teeth, or what remains of them. listen closely, you will hear us growl; we are becoming crones, we care more for serenity and less for what others think. with less to lose we are reckless and daring and when that happens, will you miss us sneaking up behind you?

The Many Timekeepers
© Diane Norrie 2016

© Mari Susan Selby 2017

X. EMBER GLOW
Moon X: September 28–October 27
New Moon in ♎ Libra Sept. 28; Full Moon in ♈ Aries Oct. 13; Sun in ♏ Scorpio Oct. 23

Hiding in Plein View © *Karen Russo 2017*

September / October

septiembre / octubre

© Tarra "Lu" Louis-Charles 2016

African Queen

♎︎
♏︎

Monday
30

ᗞ→♏︎   2:42 am
ᗞ☍♅   11:56 am
♀□♇   9:18 pm

♏︎

Tuesday
1

October

ᗞ⚹♄   1:43 am
ᗞ△♆   6:12 am
ᗞ⚹♇   12:45 pm

♏︎
♐︎

Wednesday
2

ᗞ⚹♂   2:46 am v/c
ᗞ→♐︎   4:44 am
☉⚹ᗞ   9:42 pm
♇D   11:39 pm

♐︎

Thursday
3

☿→♏︎   1:14 am
ᗞ□♆   10:14 am
ᗞ☌♃   1:40 pm
♂→♎︎   9:22 pm

♐︎
♑︎

Friday
4

ᗞ⚹♀   12:34 am v/c
ᗞ→♑︎   10:43 am
ᗞ□♂   11:26 am
ᗞ⚹☿   2:52 pm
ᗞ△♅   9:07 pm

ALL ASPECTS IN PACIFIC DAYLIGHT TIME; ADD 3 HOURS FOR EDT; ADD 7 HOURS FOR GMT

She Wants Red

Maiden, Mother, Queen, Crone □ RM Allen 2017

She wants red
a ruby slick of paint
on her brush

not a youthful cherry red—
an older woman's dark lipstick,
wine red eyeshadow,
garnet drop earrings
glinting like pomegranate seeds.

She wants red? now?
when the last of her moon blood
is leaving break-up letters
on her underwear

flash of a hummingbird throat

She wants red.
she licks the blush
from a rose hip, whispers *yes*

feels her body's red places redden
she's all smoldering coals, she

makes borscht
beet juice stains her hands
she wants to fill the tub with it
and soak

she orders red flannel sheets
queen-size for her solitary bed
sleeps in a glowing cave

dreams of a room in her house
she's never entered
on the far wall a strange woman
is carving the word
change

© *Sophia Rosenberg 2015*

ᚻᚻᚻ sábado

♑

Saturday
5

♂⚹♃	9:05 am
☉□☽	9:47 am
☽♂♄	1:39 pm
☽⚹♆	6:26 pm

Waxing Half Moon in ♑ Capricorn 9:47 am PDT

☉☉☉ domingo

♑
≈≈

Sunday
6

☽♂♇	2:15 am
☽□♀	4:25 pm v/c
☽→≈	8:42 pm
♂☍♅	11:17 pm

October
Shí yuè

Sunphant © Tamara Phillips 2013

─── ☽☽☽ xīng qī yī ───

 ≈

Monday
7

☽△♂ 12:43 am
☽□♅ 7:31 am
☽□♀ 8:36 am
☉□♄ 12:07 pm

─── ♂♂♂ xīng qī èr ───

≈

Tuesday
8

☉△☽ 2:20 am
♀→♏ 10:06 am
☽✶♃ 11:26 am v/c
♂⚼♇ 5:33 pm

Time to order We'Moon 2020!
Free Shipping within the US October 8-13th! Promo Code: Lucky13 www.wemoon.ws

─── ☿☿☿ xīng qī sān ───

≈
♓

Wednesday
9

☽→♓ 9:05 am
☽△♀ 11:47 am
☽✶♅ 7:55 pm
☉⚼♆ 8:24 pm

─── ♃♃♃ xīng qī sì ───

♓

Thursday
10

☽△♀ 4:48 am
☽ApG 11:30 am
☽✶♄ 2:08 pm
☽♂♆ 6:33 pm
♀⚼♇ 8:31 pm

─── ♀♀♀ xīng qī wǔ ───

♓
♈

Friday
11

☽□♃ 1:02 am
☽✶♇ 2:55 am v/c
☽→♈ 9:46 pm

─────────────────────

ALL ASPECTS IN PACIFIC DAYLIGHT TIME; ADD 3 HOURS FOR EDT; ADD 7 HOURS FOR GMT

The Honors of Age

When I am old
I will be an elephant
The wrinkles on my skin
Will be
The only known recording of
My Truth
Like the bark on the trees
In the ancient
Old growth forests
I will have valleys and ravines
Stretched into my surface
Marking the places where
Abundance grows wild
And fairies come to dance
In the shade of my eyelashes
There will be
Forgiveness for everything
Everyone
I will have giant ears like Aquarius
And awesome tusks
To signify my righteous place in the kingdom of the universe
When I trumpet the vibrations are tuning to the Earth
Rhythms that carry my feet Back to the water
When I am old

© Amy A. Ehn 2017

Divine Power
¤ *Sundara Fawn 2008*

ꑂꑂꑂ xīng qī liù ———

♈ ◯ Saturday
12

♂⊼♅ 1:00 am
☽☌♂ 8:43 am
♀⚹♅ 3:07 pm

⊙⊙⊙ lǐ bài rì ———

♈ ◯ Sunday
13

☽□♄ 2:36 am   ⊙☍☽ 2:08 pm
⊙⚹♃ 11:02 am ☽□♇ 2:59 pm v/c
☽△♃ 1:55 pm   ☿⚹♄ 11:56 pm

Full Moon in ♈Aries 2:08 pm PDT

October
Ayu—Fruit Month

───── ꓄꓄꓄ Henesháal—East Day ─────

♈
♉

Monday
14

⊙□♇ 12:38 am
☽→♉ 9:24 am
☽♂♅ 7:29 pm

Arise © *Jenny Hahn 2014*

───── ♂♂♂ Honesháal—West Day ─────

♉

Tuesday
15

☽☌♀ 1:33 am
☽△♄ 1:44 pm
☿△♆ 3:44 pm
☽⚹♆ 5:23 pm
☽☌♉ 5:33 pm

───── ☿☿☿ Hunesháal—North Day ─────

♉
♊

Wednesday
16

☽△♇ 1:37 am v/c
☽→♊ 7:30 pm

───── ♃♃♃ Hanesháal—South Day ─────

♊

Thursday
17

☽△♂ 12:11 pm

───── ♀♀♀ Rayilesháal—Above Day ─────

♊

Friday
18

☽□♆ 2:25 am
☽☌♃ 10:58 am
⊙△☽ 7:14 pm v/c

───────────────────────────────

ALL ASPECTS IN PACIFIC DAYLIGHT TIME; ADD 3 HOURS FOR EDT; ADD 7 HOURS FOR GMT

Goddess Speed

May you flow,
like a sweltering red hot
body of lava
traveling toward the sea.

May you shift with ease
as you encounter
that which would
stop you,
or that which would
change you,
and be changed,
like a hot liquid mass
becoming the solid earth
it longs for.

□ *Jennifer Lothrigel 2017*

Down to the Sea © *Shoshanah Dubliner 2008*

──────── ♄♄♄ Yileshàal — Below Day ────────

♊
♋

Saturday
19

☽→♋ 3:43 am
☽⚹♅ 12:47 pm
☿⚹♇ 3:21 pm
☽□♂ 10:35 pm

──────── ☉☉☉ Hathameshàal — Center Day ────────

♋

Sunday
20

☽△♀ 6:26 am
☽⚼♄ 6:29 am
♀⚹♄ 6:58 am

☽△♆ 9:17 am
☽⚼♇ 5:07 pm
☽△☿ 7:06 pm

─────────────────────────────

October
aktober

© Tamara Phillips 2015

Timber

♋
♌

Monday
21

⊙□☽   5:39 am v/c
☽→♌   9:28 am
♀△Ψ 12:40 pm
☽□♅   5:56 pm

Waning Half Moon in ♋ Cancer 5:39 am PDT

♌

Tuesday
22

☽⚹♂   5:58 am
☽□♀   3:54 pm
☽△♃ 10:41 pm

♌
♍

Wednesday
23

☽□☿   2:14 am v/c
⊙→♏ 10:20 am
☽→♍ 12:29 pm
⊙⚹☽ 12:39 pm
☽△♅   8:23 pm

Sun in ♏ Scorpio 10:20 am PDT

♍

Thursday
24

☽△♄   1:05 pm
☽☍Ψ   3:07 pm
☽⚹♀ 10:02 pm
☽△♇ 10:26 pm

♍
♎

Friday
25

☽□♃ 12:37 am
♀⚹♇   2:52 am
☽⚹☿   5:59 am v/c
☽→♎   1:20 pm
⊙⚼♄   8:29 pm

ALL ASPECTS IN PACIFIC DAYLIGHT TIME; ADD 3 HOURS FOR EDT; ADD 7 HOURS FOR GMT

2019 Year at a Glance for ♏ Scorpio (Oct. 23–Nov. 22)

In the beginning of 2019, you may be feeling anxious and riled up. Release the stored-up energy in a process of physical healing rather than trying to be productive with it. Channel your feelings using art, imagery, and symbols to gain insight into what's stressing you out.

This year, Jupiter brings a shift of consciousness to what you value. New resources—whether material or energetic—bring the potential to create greater security. Learn to accept the generosity of the universe and how blessings may come in unusual ways. On the spiritual level, you can examine your beliefs about money and resources, releasing binary labels like "success" and "failure."

Through Saturn and Pluto's guidance, you're learning about boundaries in communication, like asking for what you need and saying no to unfair demands. This kind of "Holding Action" is vital to diminishing harm and making space for positive discourse in the wider world. On a deeper level, there are stories you tell about yourself and the world that have become too rigid. It's time to generate some new stories—perhaps ones in which the world is safe, nourishing, and sensitive to your needs.

Spring brings fresh relationships that inspire wonder and enthusiasm, even with people that aren't "your type." Embrace it! These unlikely folks may catalyze new sparks of sensuality and pleasure in your life. This Uranian influence can guide you to more sustainable and fulfilling relationships.

Rhea Wolf © Mother Tongue Ink 2018

Wounded Healer
© Emily Kell 2017

──── ♄♄♄ Šanbe ────

♎

Saturday
26

)PrG 3:37 am
)♂♂ 12:48 pm
)□♄ 1:32 pm
)□ℙ 10:38 pm

──── ☉☉☉ Yekšanbe ────

♎
♏

Sunday
27

)⚹♃ 1:22 am v/c
♂□♄ 7:31 am
)→♏ 1:29 pm

♅PrH 1:31 pm
☉♂) 8:38 pm
)☍♅ 8:58 pm

Lunar Samhain

New Moon in ♏ Scorpio 8:38 pm PDT

TimeKeeper

The sill is wide;
no glass, no curtain.
At the opening, I perch—
trembling and so curious.
Looking out the window,

Looking *in* the window.
Sitting on Death's sill,
these living days I keep the time.
Lolling on the ledge,
every exhaled breath, I practice

leaving the body's sweetness.
My days roll on,
but Death can call me anytime
to jump from my adventure
under the Timekeeper Sun,

to travel to the next.
Preparations? None,
for there's no itinerary,
and the journey requires only
a single change of garment

that shall be handed me.
So these living days
I keep the time,
sitting delighted
upon the sill.
But then,
I'll let the time run out
and join my sun within
to hers.

¤ *Susa Silvermarie 2017*

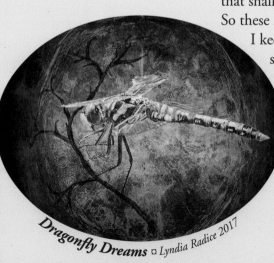

Dragonfly Dreams ¤ *Lyndia Radice 2017*

Gracias a La Vida © *Nancy Watterson 2017*

Grandmother

When Grandma was dying, we moved her bed out of doors to the garden, under the edge of the green cathedrals boughs—The place where she could see the face of God above her. She lay still for a long time, just looking up—and then, almost silently whispered, "Thank you."

All she left unfinished, now lies in my lap. I release the mantle of her sorrow, and we both are freed. I have only one wish left—may my last words be, "thank you."

excerpt ¤ Marianna Jones 2017

October / November
octubre / noviembre

Release Your Story

⊅⊅⊅ lunes

♏

Monday
28

⊙☌♅ 1:15 am
☽⚹♄ 2:19 pm
☽△♆ 3:52 pm
♂⚼♆ 8:57 pm
☽⚹♇ 11:32 pm

♂♂♂ martes

♏
♐

Tuesday
29

☽☌♀ 8:14 am
☽☌♅ 10:34 am v/c
☽→♐ 2:58 pm

☿☿☿ miércoles

♐

Wednesday
30

☿☌♀ 3:05 pm
☽□♆ 6:49 pm
☽⚹♂ 9:10 pm

♃♃♃ jueves

♐
♑

Thursday
31

Samhain / Hallowmas

☽☌♃ 7:29 am v/c
☿R 8:41 am
☽→♑ 7:38 pm

♀♀♀ viernes

♑

Friday
1

November

☽△♅ 3:46 am
⊙⚹☽ 12:21 pm
♀→♐ 1:25 pm

ALL ASPECTS IN PACIFIC DAYLIGHT TIME; ADD 3 HOURS FOR EDT; ADD 7 HOURS FOR GMT

Hallowmas

In the mirror of the Wheel, a woman of authority faces us. Her face changes and changes as we scry. When it settles, we look deep and admit everything. We let deep quiet uncover the power of our vulnerability. We listen for messages in the song of the fire, on the winds, in the sound of the river. We celebrate the unseen, and the non-doing nakedness of winter. We offer to the fire everything that is finished in our life from the old year.

The veil between the worlds is nearly sheer. This is the time to join ceremonially with our ancestors, whose energy dances down through our genes, and slides to coincide in us. We honor and embrace the wealth inherited from them.

Hallowmas is the witches' New Year's Eve. Giddy, invoking the foremothers to lend us power, we celebrate in that endless ring called a circle. Urgent and earnest, we muscle potent magik into real world manifestation. *Susa Silvermarie © Mother Tongue Ink 2018*

Crone of Flames ~ Cerridwen © Daughters of the Moon Tarot 1991

Alice's Bear Hills II © *Carol Wylie* 2003

When a Wise Woman Passes

When a wise woman passes . . .
We are left to spread
her wisdom like seeds

excerpt ¤ Sisterdiscordia 2016

Forest

When you are in a forest—think of all the levels. Down deep, where you can't see, is a whole world. Roots coil and curve in every direction . . . worms, mud, bugs and seeds among rock, and below, the molten zone.

But here on the ground is the richest realm of all, with its soil mulched by the deaths of the countless, and new life springing ever upward into brilliant flowering bulb and bush, tree and bromeliad. Birds and millions of years of animals roam, woman and man and children resonating like music with the leaves and moist air. Sunlight streaming down through the sapphire blue light—clouds drifting in perpetual journey—and above them, hundreds of thousands of galaxies, star diamonds birthing and dying.

We humans are like the forest with our root people, worm people, mud people, seed people, orchid people, short yews and towering sequoias, and a cosmos of animal people and flying bird people—above, the airy sphere of cloud people, blazing sun people, soft moon people and the innumerable star people who sparkle brilliantly in the blackest night. And all, evolving deeper and wider and higher into the unknowable Mystery.

¤ Janine Canan 2016

——— ♄♄♄ sábado ———

♑

Saturday
2

☽☌♄ 12:29 am
☽⚹♆ 1:35 am
☽□♂ 7:11 am
☽☌♇ 10:39 am
☽⚹☿ 10:46 pm v/c

——— ☉☉☉ domingo ———

♑
♒

Sunday
3

☽→♒ 3:19 am
♀△♃ 5:20 am
☽⚹♀ 7:42 am
☽□♅ 11:50 am

Daylight Saving Time Ends 2:00 am PDT

November

shí yī yuè

First Crescent

▷▷▷ xīng qī yī

Monday
4

⊙□☽ 2:23 am
☽△♂ 8:28 pm
♀⊼♅ 11:09 pm

Waxing Half Moon in ≈ Aquarius 2:23 am PST

♂♂♂ xīng qī èr

Tuesday
5

♂□♇ 2:28 am
☽⚹♃ 3:28 am
☽□☿ 6:37 am v/c
☽→♓ 3:08 pm
☽⚹♅ 11:43 pm

☿☿☿ xīng qī sān

Wednesday
6

☽□♀ 2:41 am
⊙△☽ 8:25 pm
☽⚹♄ 11:18 pm
☽♂♆ 11:37 pm

♃♃♃ xīng qī sì

Thursday
7

☽ApG 12:34 am
☽⚹♇ 9:32 am
☽△♅ 2:52 pm
☽□♃ 5:12 pm v/c

♀♀♀ xīng qī wǔ

Friday
8

☽→♈ 3:48 am
⊙⚹♄ 9:06 am
⊙△♆ 9:56 am
♄⚹♆ 6:45 pm
☽△♀ 10:15 pm

ALL ASPECTS IN PACIFIC STANDARD TIME; ADD 3 HOURS FOR EST; ADD 8 HOURS FOR GMT

Seed of Hope

We begin the true year's cycle in darkness
As tiny seeds (how we shall grow!)
But now we are potent, unformed,
Deep dreaming in the earth below.
The embers of the fires do not stop burning
Always, always we carry them as we go.

This is the time to fill your cup, my sisters,
Fill your cup ready for what's to come . . .
We have pressed rose red apples,
Gifts to sweeten your senses
And carry you onward into the dark
Where we all must go.

© Denise Ostler 2017

So each of you fill your cup, dear sisters, in tender care for body and spirit. Know even a moment spent in deep resting silence will set spirit dancing long with delight. Fill your cup full of nourishment for the cycle ahead. In the stillness of the circle, as all realms draw near, dance a welcome for your ancestors. Let your heart hear them say: You are so blessed by all who came before.

I am a seed of dreaming, with a heritage of thousands;
I will choose well how I grow.

© Nell Aurelia 2016

———— ♄♄♄ xīng qī liù ————

♈ ☽ **Saturday**
9

☽□♄ 11:55 am
☿⚹♇ 6:09 pm
☽□♇ 9:37 pm

———— ☉☉☉ lǐ bài rì ————

♈ ☽ **Sunday**
♉   **10**

☽☍♂ 3:58 am
☽△♃ 6:00 am v/c
☿PrH 11:08 am
☽→♉ 3:18 pm
☽♂♅ 11:10 pm

November
Athon—Seed Month

© Sarah Cook 2016

A Tea Offering

—————— ◗◗◗ Henesháal—East Day ——————

♉

Monday
11

☉☌♉ 7:22 am
☽⚹♆ 10:11 pm
☽△♄ 10:44 pm

Mercury Transits Sun in ♏ Scorpio 4:34 am PST*

—————— ♂♂♂ Honesháal—West Day ——————

♉

Tuesday
12

☽☍♉ 1:51 am
☉☍☽ 5:34 am
☽△♇ 7:48 am v/c
♂⚹♃ 10:21 am

Full Moon in ♉ Taurus 5:34 am PST

—————— ☿☿☿ Hunesháal—North Day ——————

♉
♊

Wednesday
13

☽→♊ 12:46 am
☿⚹♄ 6:35 am
☉⚹♇ 10:00 am
☿△♆ 2:34 pm

—————— ♃♃♃ Hanesháal—South Day ——————

♊

Thursday
14

☽☍♀ 6:16 am
☽□♆ 6:32 am
♀□♆ 9:06 am

—————— ♀♀♀ Rayilesháal—Above Day ——————

♊
♋

Friday
15

☽☍♃ 1:24 am
☽△♂ 3:40 am v/c
☽→♋ 8:15 am
☽⚹♅ 3:16 pm

—————————————————————————————

*Transit visible over SW Europe, SW Asia, Africa, Americas, Pacific, Atlantic, I. Ocean, Antartica

Confession

Pieces of me are scattered all over this place. I lost a dream on that corner once, it was a nice one. And over there my cat died, first slowly and then with alarming speed. One block from here a favorite childhood climbing tree was cut down by a city parks worker; I guess it was sick, or maybe blocking a power line. I turned my back on compassion when I learned about sink or swim, survival of the fittest, show no mercy. In that schoolyard I surrendered community. I had to use 80 layers of concrete mud to seal that vault. It kept flying open asking me to play.

But I found some things too—I found guilt in that church, the charming one on the pond, and isolation when we left it behind. I picked up manipulation about that time, in the hallways, on the busses. I learned the skills of desire, control, defeat, despair. On that rooftop I found solace and oblivion in a stolen bottle of wine, and communion in the rhythm of puff puff pass. Then I lost it all again.

So now begins the process of picking up and dusting off, of putting back and trying on—of setting aside and reclaiming. Lost pieces of me coming home and false fronts fading slowly, slipping skins and molting feathers—growing lush new plumage, learning what really makes me smile. Remembering my feet were made for dancing, not running away, and my voice for singing not screaming in the dark, or at least not screaming quite so much. A time now of returning—wholeness filling the hole, home healing the homesick, love relieving the loss. The scattered unscattering.

¤ *Marianna Jones 2015*

ꜩꜩꜩ Yìleshǎal—Below Day

♋ 🌒 Saturday
16

☽△♅ 8:08 am
☽△Ψ 1:02 pm
☽☍♄ 2:18 pm
☽☍♇ 10:15 pm

☉☉☉ Hathameshǎal—Center Day

♋ 🌒
♌ Sunday
17

☉△☽ 4:53 am
☽□♂ 12:14 pm v/c
☽→♌ 1:57 pm
☽□♅ 8:35 pm

November
novamber

♌ ## Monday
18

D□☿ 10:53 am
♂→♏ 11:40 pm

♌
♍ ## Tuesday
19

D△♀ 4:04 am
D△♃ 1:06 pm
⊙□D 1:11 pm v/c
D→♍ 5:54 pm
D✶♂ 6:48 pm

Waning Half Moon in ♌ Leo 1:11 pm PST

♍ ## Wednesday
20

D△♅ 12:12 am
♅D 11:12 am
D✶♅ 1:32 pm
D☍♆ 8:51 pm
D△♄ 10:43 pm

♍
♎ ## Thursday
21

D△♇ 5:42 am
♂⅄♄ 11:31 am
D□♀ 11:39 am
D□♃ 4:25 pm
⊙✶D 7:31 pm v/c
D→♎ 8:19 pm

♎ ## Friday
22

⊙→♐ 6:59 am
DPrG 11:36 pm

Sun in ♐ Sagittarius 6:59 am PST

ALL ASPECTS IN PACIFIC STANDARD TIME; ADD 3 HOURS FOR EST; ADD 8 HOURS FOR GMT

2019 Year at a Glance for ♐ Sagittarius (Nov. 22–Dec. 21)

Who's *that*? The question implies attraction, intrigue, curiosity. In 2019, greet yourself with this question whenever you look in the mirror. The revolution of love comes to you via Jupiter's transit through your home sign. The planet of meaning-making and abundance wants you to know that your identity is ready for rebirth. The end of one chapter initiates the beginning of the next. Reflect on the themes of the last 12 years, and dream about what comes next. If you respond only to outer world events, you'll get trapped in a cycle of overwhelm and retreat. Instead, locate the burning desire at the core of you. Let that be your guide.

Personal resources—including material possessions and values—will be getting a major overhaul this year, thanks to Saturn and Pluto. If you're stuck in some old stories about money and property, let them go and figure out what you actually want. You may experience scarcity, but you'll also feel empowered to re-focus your priorities and simplify lifestyle choices accordingly.

2019 will also provide motivation to renovate your health and daily rituals. Uranus says it's time to get creative about caring for yourself. Listen to the Earth and your own body as sources of wisdom to create structures that sustain wellness in your life. Share your innovative solutions with others. Come December, get ready to slow down and integrate the illuminating insights and hope you've gathered for the future.

Rhea Wolf © Mother Tongue Ink 2018

Ephemeral Focus
© Autumn Skye ART 2015

───────── ꌇꌇꌇ Šanbe ─────────

♎︎
♏︎

Saturday
23

☽□♄	12:53 am	
☽□♇	7:32 am	
☽⚹♀	6:00 pm	
☽⚹♃	6:49 pm v/c	
☉△⚷	8:55 pm	
☽→♏︎	9:58 pm	

───────── ☉☉☉ Yekšanbe ─────────

♏︎

Sunday
24

☽♂♂	3:36 am
☽☍♅	3:51 am
♀♂♃	5:33 am
♂☍♅	8:51 am
☽♂♀	7:50 pm

ask the wind

i want my country run
by native intelligence

i want every development decision
run past sister otter & brother pine
i want to elect someone
who knows you have to ask & thank & honor
oak before you take acorns
someone who knows the best use of wetlands
is wet

someone who will never find a good reason
to poison
air or water or soil
who's kin to even pugnacious weeds

i want my government managed
by men & women who ask the wind
which way to blow
who put their hand on the trunk
of redwood to feel their own heart beat

i want all the laws with no kindness in them
made into smoke signals

when i go to the websites
  of native environmentalists
    i feel a cool wind that smells
    good

      i want to smell that wind
      every time
      i read the newspaper
      i want to smell that wind
      in public halls

□ *Sandy Eastoak 2008*

© *Evangeline Coomber 2000*

XII. GLOBAL COMMUNITY
Moon XII: November 26–December 25
New Moon in ♐ Sagittarius Nov. 26; Full Moon in ♊ Gemini Dec. 11; Sun in ♑ Capricorn Dec. 21

Dark Spirits Roam Free © Glenda Goodrich 2015

November / December
noviembre / diciembre

© Melissa Renteria 2015

Monday
25

☽△♆	12:27 am
☽⚹♄	2:59 am
☽⚹♇	9:30 am v/c
♀→♑	4:28 pm
☉⚼♅	5:59 pm

Roots Run Deep

Tuesday
26

☽→♐	12:11 am
☉☌☽	7:06 am
♀□♃	10:28 pm

New Moon in ♐ Sagittarius 7:06 am PST

Wednesday
27

☽□♆	3:37 am
♆D	4:32 am

Thursday
28

☿△♆	1:51 am
☽☌♃	2:50 am v/c
☽→♑	4:33 am
♀△♅	10:27 am
☽△♅	10:41 am
☽☌♀	10:43 am
☽⚹♂	4:06 pm

Friday
29

☽⚹♆	9:39 am
☽⚹♅	12:30 pm
☽☌♄	1:17 pm
☽☌♇	7:57 pm v/c
☿⚹♄	10:12 pm

ALL ASPECTS IN PACIFIC STANDARD TIME; ADD 3 HOURS FOR EST; ADD 8 HOURS FOR GMT

At the Heart of it

Somewhere beyond the map,
we will excavate
the ancient meeting tree.
The bridge between worlds;
it's the intersection of layers
sewn like cloth upon cloth.
Its branches & roots are
formed of our own bones.
Together we are the stitches
that link the layers.
Gazing through the gaps
between the fabric,
our eyes
are the clearest
they've ever been.

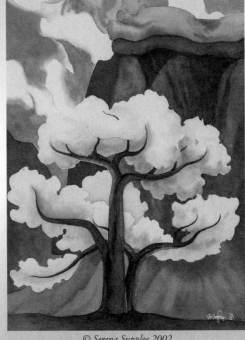

© Serena Supplee 2002

There are no boundaries;
anymore

excerpt ¤ Amanda Heinz-Stevenson 2017

ᎭᎭᎭ sábado

♑
♒

Saturday
30

☽→♒ 12:13 pm
☽□♅ 6:38 pm

☉☉☉ domingo

♒

Sunday
1

December

☽□♂ 3:43 am
☉⚹☽ 5:43 am

December
shí èr yuè

IsaIsa © *Dorrie Joy 2017*

 ☽☽☽ xīng qī yī

♒
♓

Monday
2

☽□♉ 4:27 am v/c
♃→♑ 10:20 am
☿⚹♇ 9:22 pm
☽→♓ 11:10 pm
☽⚹♃ 11:25 pm

────── ♂♂♂ xīng qī èr ──────

♓

Tuesday
3

☽⚹♅ 5:43 am
♀⚹♂ 7:47 am
☽△♂ 6:50 pm
☽⚹♀ 7:26 pm
☉□☽ 10:58 pm

Waxing Half Moon in ♓ Pisces 10:58 pm PST

────── ☿☿☿ xīng qī sān ──────

♓

Wednesday
4

☽♂♆ 7:19 am
☽⚹♄ 12:14 pm
☽⚹♇ 6:41 pm
☽ApG 8:01 pm

────── ♃♃♃ xīng qī sì ──────

♓
♈

Thursday
5

☽△♉ 12:14 am v/c
☽→♈ 11:44 am
☽□♃ 1:09 pm

────── ♀♀♀ xīng qī wǔ ──────

♈

Friday
6

☽□♀ 2:57 pm
☉△☽ 5:04 pm

ALL ASPECTS IN PACIFIC STANDARD TIME; ADD 3 HOURS FOR EST; ADD 8 HOURS FOR GMT

Ramblings of an Old Old Soul

I was on Syilx land picking poplar buds one day, when a curious woman walked by and asked what I was doing. I told her I use the buds to create a pain-relieving balm. She wanted to know how I learned about it. "From my Medicine People," I said, thinking of my witchy friends and siStars, who have shared earthly knowledge with me. "You have such a neat culture," she said. I stared blankly, thinking, which culture is that? She continued on her way, and I began to ponder . . . I've lived too many lives to be this or that: a woman, a man, red, yellow, white, or black. I've lived in all four corners of the world, and I remember. I have lived as a slave, a queen, a healer, a hunter, a peasant. I know myself beyond this lifetime, beyond experience, beyond culture, race, profession, status. I am not this or that—I am both this *and* that.

People often ask, "Where are you from?" seeking an explanation for the color of my skin and kink in my hair. The answer, London, Ontario, never satisfies. They dig until my Jamaican heritage emerges. Wherever I go, people see a different culture in me. In Greece, I'm Greek; in Spain, Spanish; and in the Caribbean, they see my ancestors in me. I am as the Jamaican motto states—*Out of Many One People.* I see myself in faces from around the globe. Ego sees falsely the separation of color and gender. Soul and Spirit are color blind, gender-neutral. I found myself picking poplar buds, dreaming of a world where ego is transparent, and Spirit shines through—a Soul filled world, where Humanity is Our Culture.

© *Mahada Thomas 2015*

ㅅㅅㅅ xīng qī liù

♈ ☽ ### Saturday
♉   7

☽□♄   1:05 am
☽□♇   7:01 am v/c
☽→♉ 11:29 pm

☉☉☉ lǐ bài rì

♉ ☽ ### Sunday
       8

☉□♆   1:00 am
☽△♃   1:58 am
☽☌♅   5:34 am
♀✶♆   1:48 pm
♃□♎   8:27 pm

December
Adol—Root Month

Moon Magic © Sue Davis 2016

───── ꓛꓛꓛ Henesháal—East Day ─────

♉

Monday
9

☽☍♂ 1:10 am
☿→♐ 1:41 am
☽⚹♆ 6:20 am
☽△♀ 8:07 am
☽△♄ 11:54 am
☽△♇ 5:13 pm v/c

───── ♂♂♂ Honesháal—West Day ─────

♉
♊

Tuesday
10

♉△☿ 1:32 am
☽→♊ 8:47 am
☽☍☿ 12:43 pm

───── ☿☿☿ Hunesháal—North Day ─────

♊

Wednesday
11

♀☌♄ 2:05 am
☿⚼♅ 3:54 am
☽□♆ 2:11 pm
☉☍☽ 9:12 pm v/c

───── ♄♄♄ Hanesháal—South Day ─────   **Full Moon in ♊ Gemini 9:12 pm PST**

♊
♋

Thursday
12

☽→♋ 3:23 pm
☿D   6:28 pm
☽☍♃ 7:33 pm
☽⚹♅ 8:43 pm

───── ♀♀♀ Rayilesháal—Above Day ─────

♋

Friday
13

♂△♆ 3:54 am
♀☌♇ 7:16 am
☽△♆ 7:37 pm
☽△♂ 8:25 pm

Borders

Araceli informs me we are in a border town, and the other side of the fence is full of dangerous people. Drug people. Lost people. Coyote people. She overheard this information when the news was blaring out of the television in the pet hospital. Pets shouldn't have to be involved in the politics of people. Pom Pom isn't a revolutionary for any political movement, but I know he understands the purity of loving a grandma. *What would the news look like if dogs were in charge? It would be a god damn spectacular miracle.* I imagine Abuelita exclaiming this and slamming her fist on any flat surface in the Mango House.

I ask Araceli if she has ever seen the fence lining the curving roads when Mami drives us into the city. My little sister nods and I tell her, there are beautiful souls on the other side of the fence. Some of them are lost, sure, but the fire inside your blood which makes you love ferociously is inside the bodies of people over there, too. There are babies and eloteros and mujeres who sell those amazing neon light-up toys on that side of the fence. There are abuelas and granddaughters and ladies who work in skyscrapers, with perfect hair and fancy nails on that side of the fence. There are forces of good and the essence of evil on all sides. Abuelita was born on that side of the fence. Our ancestors lived and thrived on that side of the fence. We are from here, but part of us is also on that side of an imaginary line.

from Itzá *published by Broken River Books © Rios de la Luz 2017*

ᖾᖾᖾ Yilesháal—Below Day

♋ ♌ ⬤ ## Saturday
14

☽☌♄ 1:32 am
☽☍♇ 5:47 am
☽☍♀ 7:56 am v/c
☽→♌ 7:56 pm

⊙⊙⊙ Hathamesháal—Center Day

♌ ◐ ## Sunday
15

☽□♅ 1:02 am
♃△♅ 11:01 am
☽△☿ 12:18 pm

Winter Solstice

Observing the cycle of changing light on our dear planet all year long has brought us to Winter Solstice, the simultaneous ending and beginning of the solar year. A blessed poise suspends us.

The Great Mother gives birth to the first speck of reborn light, barely believable after its agonizing absence. The longest night and the start of increase was originally called Yule, from the Anglo-Saxon Iul, wheel. At the center of the Wheel we look in the mirror. We see the seeds of our intentions, our hopes for Gaia's future.

She exhorts us: Maids, Make Merry! Burning only seasoned wood, we know it represents our now experienced selves. Around the Yule fire, we connect with the particular land upon which we stand, and we send out benedictions and protections for that place. In the deep dark, we join our firecircle with all the circles celebrating on earth. Into the planet's chalice of joy, we pour our own, and we see it overflow for all.

Susa Silvermarie © Mother Tongue Ink 2018

Awakening ¤ Joan Zehnder 2016

Ambrosia © *Autumn Skye ART 2012*

Solstice flame,
remnant of First Light,
daughter of dancing suns,
you flicker with primordial fire.

Sister of the Wellspring,
imbue with peace this quiet place,
soften our ragged edges
and beckon to your light
those who suffer and are lost.

Prancing flame,
kin to all who labor to live,
breathe into us the radiance of your fire
so we will remember
the light from which we come.

May this Day of Turning renew us in midwinter
so spring will find in us a fertile field
where kindly things can thrive, multiply
and walk with generosity!

© *Pam Ballingham 2011*

December

desamber

Bast the Cat Goddess

© Emily Balivet 2011

─── ☽☽☽ Dŏsanbe ───

♌︎
♍︎

Monday
16

☽□♂ 2:46 am
☉△☽ 2:10 pm v/c
☽→♍︎ 11:16 pm

─── ♂♂♂ Sešanbe ───

♍︎

Tuesday
17

☽△♅ 4:12 am
☽△♃ 4:56 am
☽□☿ 9:29 pm

─── ☿☿☿ Cahâršanbe ───

♍︎

Wednesday
18

☽☍♆ 2:29 am
☽⚹♂ 8:14 am
☽△♄ 8:56 am
☽PrG 12:17 pm
☽△♇ 12:29 pm
☉□☽ 8:57 pm

Waning Half Moon in ♍︎ Virgo 8:57 pm PST

─── ♃♃♃ Panjšanbe ───

♍︎
♎︎

Thursday
19

☽△♀ 12:07 am v/c
♂⚹♄ 2:00 am
☽→♎︎ 2:04 am
☽□♃ 8:34 am
☿□♆ 8:19 pm
♀→♒︎ 10:41 pm

─── ♀♀♀ Jom'e ───

♎︎

Friday
20

☽⚹♅ 6:22 am
☽□♄ 12:08 pm
☽□♇ 3:24 pm

ALL ASPECTS IN PACIFIC STANDARD TIME; ADD 3 HOURS FOR EST; ADD 8 HOURS FOR GMT

2019 Year at a Glance for ♑ Capricorn (Dec. 21–Jan. 20)

I know that you put a lot of pressure on yourself, Capricorn. Which is exactly why I encourage you to return to a sense of wonder in 2019. You may harrumph this idea initially. You may think that your responsibilities can't wait for an epiphany. But if you can tune into the miracles of life, Jupiter will shift your consciousness by exposing profound truths that allow you to be even more effective in offering your skills to the world. To align with this energy, take a spiritual or healing retreat this year, cultivate a sense of empathy by writing letters to those who are incarcerated, or simply get on your knees and pray!

The eclipses in January and July prompt you to make changes in close relationships. You might need to adjust communication and practice self-control rather than try to control others. Listen to what your loved ones need—it could be very different from what you thought.

Pluto and Capricorn are releasing you from old identities, teaching you to be more vulnerable and sensitive with your partners, friends, and community. It will feel unfamiliar and uncomfortable, but ultimately lead to healthier and more satisfying bonds.

Uranus in Taurus invites you to rekindle a practice that makes you feel like a kid again. Spend time gardening, rock-hunting, or working with textiles to illuminate new creative purpose. When you pay attention to children or invite romance into your world, you will discover unrecognized resources that foster sustainable growth.

Rhea Wolf © Mother Tongue Ink 2018

— ኅኅኅ Šanbe —

Saturday
21

♀✶♄ 3:14 am
☉✶☽ 3:45 am v/c
☽→♏ 4:57 am
☽□♀ 7:51 am

☽⚻♅ 9:46 am
☽✶♃ 12:20 pm
☉→♑ 8:19 am

Winter Solstice

Sun in ♑ Capricorn 8:19 pm PST

— ☉☉☉ Yekšanbe —

Sunday
22

♀□♅ 5:30 am
♂✶♇ 6:32 am
☽△♆ 8:32 am
☽✶♄ 3:51 pm

☽✶♇ 6:51 pm
☽♂♂ 7:27 pm v/c

Learning Solar Power

I stretch out my arms
and feel your warmth. All this power!
Coming so far just for us.

I am learning to live beside you
but I'm slow compared to the mountains
It's taken so long to understand grains
and pumpkins, hearthfire,
the souls of diamonds.

I've misunderstood so much—forests of oil,
sparking atoms, the worth of one burning tree.
But now I am making engines
for catching your radiance, I am honoring
water and air, altering old wrong patterns
and watching earth sciences
discover new ways to go on being
this small blue planet, last daughter
of Mother Sun.

□ Rose Flint 2017

Her Islands © *Dhira Lawrence 2012*

XIII. MOTHER SUN

Moon XIII: December 25–January 24

New Moon in ♑ Capricorn Dec. 25; Full Moon in ♋ Cancer Jan. 10; Sun in ♒ Aquarius Jan. 20

Harnessing the Wind and the Sun © *Diane Norrie 2017*

December
diciembre

─── ⊅⊅⊅ lunes ───

Monday
23

⊙□♄ 7:17 am
☽→♐ 8:34 am
☽⚹♀ 4:37 pm

─── ♂♂♂ martes ───

Tuesday
24

☽□♆ 12:56 pm
⊙△♅ 1:44 pm

─── ☿☿☿ miércoles ───

Wednesday
25

☽♂♉ 3:18 am v/c
☽→♑ 1:45 pm
♃ApH 1:54 pm
☽△♅ 6:45 pm
⊙♂☽ 9:13 pm
☽♂♃ 11:29 pm

Annular Solar Eclipse 6:29 pm PST*
New Moon in ♑ Capricorn 9:13 pm PST

─── ♃♃♃ jueves ───

Thursday
26

☽⚹♆ 7:23 pm

─── ♀♀♀ viernes ───

Friday
27

☽♂♄ 4:08 am
☽♂♇ 6:42 am
⊙♂♃ 10:25 am
☽⚹♂ 1:03 pm v/c
☽→♒ 9:20 pm

---

* Eclipse visible over E. Europe, Asia, NW Aus., E. Africa, Pacific, I. Ocean

Sun Goddess
© Rita Loyd 2010

Look What You've Done

You brought out spots and sweat on me. Made me look tanned and healthy when quite the opposite is true. You bleached the color right out of the rug in the living room, scorched those poor potted plants on the porch and the grass that was cut short. You wake me too early in the morning and linger too long on summer evenings. There's more, but I hold my tongue for fear of a fireball heading my way.

Okay. Okay. Some of it is my own fault. Our own fault for not taking better care of ourselves and planet.

So, on the other hand, I'm mesmerized by the delicate dance you do on top of the water. Am drawn to the artistry of the layers of shadows only you could create. Am amused by the cat that follows you around the house to nap. Love the colors you allow, and what you do for the moon and my crystals in the windows.

I am amazed at the life-altering tension between too much of you and not enough.

© Linda M. Evans 2017

───── ᚻᚻᚻ sábado ─────

≈

Saturday
28

ℙ□⛢   2:32 am
ℙ☌♀   6:07 pm
⛢→♑   8:55 pm

───── ☉☉☉ domingo ─────

≈

Sunday
29

⛢□♃   8:57 pm

December 2019 / January 2020

shí èr yuè / yī yuè

Monday
30

☽□♂ 2:24 am v/c
☽→♓ 7:41 am
☽⚹♀ 12:52 pm
☽⚹♅ 1:04 pm
♀△♅ 2:22 pm
☽⚹♃ 8:37 pm

Tuesday
31

☉⚹☽ 2:32 am
☽☌♆ 4:15 pm

Wednesday
1

January 2020

☽⚹♄ 2:43 am
☽⚹♇ 4:38 am
☽ApG 5:21 pm
☽△♂ 6:13 pm v/c
☽→♈ 8:00 pm

Thursday
2

♀☌♃ 8:41 am
☽□♃ 10:19 am
☽□☿ 10:32 am
☉□☽ 8:45 pm

Waxing Half Moon in ♈ Aries 8:45 pm PST

Friday
3

♂→♐ 1:37 am
☽⚹♀ 7:38 am
☽□♄ 3:49 pm
☽□♇ 5:18 pm v/c

ALL ASPECTS IN PACIFIC STANDARD TIME; ADD 3 HOURS FOR EST; ADD 8 HOURS FOR GMT

Over the Hills © *Elspeth McLean 2013*

Sun Gazing

I watch Sun sneak behind the tree line.
She dances as She sets.
She must be excited
to have a whole
new audience.

© *Anne Baggenstoss 2017*

──── ♄♄♄ xīng qī liù ────

♈
♉

Saturday
4

☽→♉   8:15 am
☽☌♅   1:31 pm
☿ApH 8:08 pm
☽△♃ 11:20 pm

──── ☉☉☉ lǐ bài rì ────

♉

Sunday
5

☽△☿   7:18 am
☉△☽   1:37 pm
♂△♄   1:53 pm
☽⚹♆   4:15 pm

WE'MOON EVOLUTION: A COMMUNITY ENDEAVOR

We'Moon is rooted in womyn's community. The datebook was originally planted as a seed in Europe where it sprouted on women's lands in the early 1980s. Transplanted to Oregon in the late '80s, it flourished as a cottage industry on We'Moon Land near Portland in the '90s and early 2000s, and now thrives in rural Southern Oregon.

The first We'Moon was created as a handwritten, pocket-size diary and handbook in Gaia Rhythms, translated into five languages, by womyn living together on land in France. It was self-published as a volunteer "labor of love" for years, mostly publicized by word-of-mouth and distributed by backpack over national borders. When We'Moon relocated to the US, it changed to a larger, more user-friendly format as we entered the computer age. Through all the technological changes of the times, we learned by doing, step by step, without much formal training. We grew into the business of publishing by the seat of our pants, starting with a little seed money that we recycled each year into printing the next year's edition. By the early '90s, when we finally sold enough copies to be able to pay for our labor, Mother Tongue Ink was incorporated as We'Moon Company, and it has grown abundantly with colorful new fruits: a datebook in full color, a wall calendar, greeting cards, a children's book, an Anthology of We'Moon Art and Writing, a goddess-poetry book, and new ventures into the field of electronic publishing.

Whew! It was always exciting, and always a lot more work than anyone ever thought it would be! We learned how to do what was needed. We met and overcame major hurdles along the way that brought us to a new level each time. Now, the publishing industry has transformed: independent distributors, women's bookstores and print-based publications are in decline. Nonetheless, We'Moon's loyal and growing customer base continues to support our unique womyn-created products, including the Anthology and the new We'Moon translation *en Español!* This home-grown publishing company is staffed by a steady and highly skilled multi-generational team—embedded in women's community—who inspire, create, produce and distribute We'Moon year in and year out.

Every year, We'Moon is created by a vast web of womyn. Our Call for Contributions goes out to thousands of women, inviting art and writing on that year's theme (see p. 236). The material is initially reviewed in Weaving Circles, where local area women give feedback. The We'Moon Creatrix then collectively selects, designs, edits, and weaves the material together in the warp and woof of natural cycles through the thirteen Moons of the year. In final production, we fine-tune through several rounds of contributor correspondence, editing and proofing. Approximately nine months after the Call goes out, the final electronic copy is sent to the printer. All the activity that goes into creating We'Moon is the inbreath; everything else we do to get it out into the world to you is the outbreath in our annual cycle. To learn more about the herstory of We'Moon, the growing circle of contributors, and the art and writing that have graced its pages over the past three and a half decades of women's empowerment, check out the anthology *In the Spirit of We'Moon* (see page 231).

Sister Organizations: We'Moon Land, the original home of the We'Moon datebook in Oregon, has been held by and for womyn since 1973. One of the first intentional womyn's land communities in Oregon, it has continued to evolve organically towards a sustainable women's community and retreat center, on 52 acres, one hour from Portland. Founded on feminist values, ecological practices and earth-based women's spirituality, we envision growing into a diverse, generationally interwoven community of friends and extended family, sharing a vision of creative spirit-centered life on the land. **We'Mooniversity** is a 501c3 tax-exempt organization that was created by We'Moon Land residents in 1999 for outreach to the larger women's community. WMU periodically co-sponsors gatherings, retreats and workshops on the land, and aspires to become a hub—online and on land—for We'Moon-related courses, projects and networking resources. **OreGaia:** Northwest Womyn's Fest, now in its 3rd year, is the newest annual event on We'Moon Land. Contact us for camping, visits and events on the land: www.wemoonland. org, www.oregaia.com, www.wemooniversity.org, www.wemoon.ws

Musawa ¤ Mother Tongue Ink 2017

WE'MOON TAROT

Wild Card
© *Jakki Moore 2013*

Announcing the We'Moon Tarot: now in process of creation! A We'Moon Tarot Deck, drawing from art and writing in We'Moon over the years, has been a subliminal work-in-progress since 1990 when we first started conceiving of We'Moon themes based on Tarot archetypes. As author/editor of this new We'Moon project, I am excited about being able to draw from We'Moon art and writing spanning the turn of this century—as an oracle for our times. We will tap into the creative wisdom of We'Moon to re-configure a Tarot Deck from a contemporary multi-cultural feminist perspective, grounded in earth-based women's spirituality. If you are interested in being part of the conversation—to review different cards and interpretations for conjuring the images that would best represent each card in a We'Moon Tarot Deck—email (wmu@wemoon.ws) with 'We'Moon Tarot' on subject line), or check out occasional postings on the **We'Moon Tarot Forum** (wemoon.ws/tarot.html).

We'Moon Themes and Tarot Correspondences: So far, we have been conjuring We'Moon themes for each year's edition based on the corresponding number of the Major Arcana cards in Tarot, which represent the cosmic forces. We explore the 4 elements of the Minor Arcana associated with the datebook themes, as well. The Minor Arcana in Tarot corresponds to the 4 elements (or suits) representing the karmic forces at work in our individual paths. While the elements of earth, water, fire and air are not named themes of particular We'Moon editions, they are naturally woven throughout the art and writing While the elements of earth, water, fire and air are not named themes of particular We'Moon editions, they are naturally woven throughout the art and writing of the latest We'Moons as follows:

• *We'Moon 2017: StarDust* (XVII): the Star, air, expanded consciousness
• *We'Moon 2018: La Luna* (XVIII): the Moon is in her element: water
• *We'Moon 2019: Fanning the Flame* (XIX): the Sun, fire is the element

And we invite our contributors to keep the remaining elements in mind when sending art and writing for upcoming editions:

• *We'Moon 2020:* (XX: the Judgement card), earth element: sustainability
• *We'Moon 2021:* (XXI: the World card), all elements join in the center

Musawa ¤ Mother Tongue Ink 2016

WE'MOON ON THE WEB

Come see what we're up to at wemoon.ws! Stay up to date with what's going on by signing up for **Weekly Lunar News**, a brief and lovely reminder of upcoming holy days, astrological and lunar events, and **We'News**, a periodic mailing announcing Mother Tongue Ink releases, specials and events! You can browse our products, old and new, keep an eye on the creation and conversation around We'Moon Tarot, and explore our astrological connection by reading the sun sign and weekly Starcodes! There's so much to learn by browsing the site—from how to donate to our Women in Prison program to exploring the history of We'Moon land and We'Mooniversity. Then you can dive into our vast web of artists and writers, get lost in spirituality and astrology, health and wellness, and even music, eco-friendly practices and green building! We hope you enjoy and come back often!

Kim Crown ¤ Mother Tongue Ink 2016

From left to right: Top Leah, Sue, Susie, Ricky
Bottom: Dana, Sequoia, Bethroot, Barb

STAFF APPRECIATION

I want to send out big kudos to the amazing women I get to work with in the We'Moon office on a daily basis. Our team worked extraordinarily hard, this year—We'Moon 2018 sold out! Sequoia, crafty and agile graphic and web designer, brought brilliant ideas to the table. Bethroot, wordwitch extraordinaire, also came out with a new book: *Preacher Woman for the Goddess: Poems, Plays, Invocations and Other Holy Writ* (see page 231). Leah, imaginative and resourceful production assistant and promo prodigy, energized us with innovative thoughts and perspectives. Sue, multitalented mastermind in production accounts and bookkeeping keeps the wheels of We'Moon oiled and lively. Susie, the many-armed goddess in the shipping department, and Dana, steadfast and hilarious shipping assistant, kept the office humming along and We'Moons flying out the door to their destinations, while keeping us laughing.

I also want to thank the talented and creative women whose work you see in these pages. You can read about each of them, starting on page 192, and become a contributor yourself! See page 236.

Barbara Dickinson © Mother Tongue Ink 2018

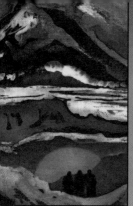

WE'MOON ANCESTORS

We honor wemoon who have gone between the worlds of life and death recently, beloved contributors to wemoon culture who continue to bless us from the other side. We appreciate receiving notice of their passing.

Barbara Blaine (1956–2017) was a dedicated activist who had been sexually abused as a teenager by a Catholic priest. She founded the Survivors' Network of Those Abused by Priests, a powerful source of advocacy and direct action. She was relentless in holding the Church accountable for protecting abusive priests.

Edith Windsor (1929–2017) was the gay rights activist whose lawsuit against the Defense of Marriage Act resulted in a 2013 Supreme Court decision to grant same-sex couples access to Federal benefits. After her lesbian partner died, Edith sued for relief from estate taxes, demanding that she be treated as a heterosexual widow would have been. She became a national celebrity.

Kate Millet (1935–2017) was a ground-breaking theoretician of 20th century feminism. Her 1970 book *Sexual Politics* was hailed as the "Bible of Women's Liberation." She called out male dominance in every aspect of life, demanding an end to patriarchy. Kate was an accomplished sculptor. She fought for abortion rights, workplace equality, sexual freedom, and wrote openly about her mental illness in *The Looney Bin Trip* (1990). She was a one-of-a-kind revolutionary icon.

Sima Wali (1957–2017) was a committed Afghan feminist who worked passionately for human rights, especially for refugees and girls fleeing from conflict. She established the Ministry of Women's Affairs, and was a founding member of the Sisterhood Is Global Institute. Simi was CEO of Refugee Women in Development. A vociferous opponent of the Taliban, she saw herself as conducting "*jihad* for social justice and peace."

Simone Veil (1928–2017) was a distinguished and dedicated feminist who fought for women's legal rights and other liberal causes in France. She was a Holocaust survivor, a lawyer and politician who focused on prison reform and disability rights as well as on feminist issues. She became the French Minister of Health, and was the first woman to become President of the European Parliament.

Ursula Le Guin (1929–2018) was a visionary feminist and literary force-of-nature who revolutionized fantasy science fiction. She imagined alternative worlds, turning our notions of power, gender, race, and religion upside down. Her popular work criticized capitalism's disregard for nature; and brought environmentalism, anarchism, radical equality into mainstream reading. She authored 20 novels; wrote poetry, essays, short stories; and made tidal waves in speculative discourse about human life and its future.

Sun Return
◻ *Debra Hall 2017*

WOMEN HUMAN RIGHTS DEFENDERS MEMORIAL

We mourn and honor here some of the WHRD who were killed in 2017 as they fanned the flames for justice. These are among the many women activists who were murdered, often in their own homes, by unknown gunmen or killed in police actions. We celebrate the extraordinary courage of these women. Thanks to the Association for the Rights of Women In Development for their online memorial tribute to the hundreds of WHRD who have died in recent years from various causes, more than half by targeted violence. (awid.org/whrd-tribute)

Colombia

Emilsen Manyoma: outspoken critic of mining companies.

Ruth Alicia López Guisao: community leader working with indigenous people on food security, health, education, agro-ecology.

Yoranis Isabel Bernal Varela: tribal leader fought for rights of indigenous women.

Idaly Castillo Narváez: community leader, member of victims' group.

Efigenia Vásquez Astudillo: radio journalist working with indigenous group, covering clashes with police.

Luz Yeni Montaño: advocate for displaced people, spoke up against systematic assassinations.

Mexico

Miroslava Breach Velducea: reporter covering organized crime and political corruption.

Miriam Rodriguez Martinez: activist and leader of Collective of Missing Persons after daughter's murder, targeted by drug cartel.

Meztli Omixochitl Sarabia Reyna: human rights defender.

Brazil

Jane Julia de Oliveira: environmental defender, rural worker advocate.

Kátia Martins: land rights leader, environmental defender, advocate for farmers.

Maria da Lurdes Fernandes Silva: community land rights leader, environmental defender, denounced illegal land seizures.

Turkey

Aysin Büyüknohutçcu: environmentalist, farmer, fought stone quarry industry.

Malta

Daphne Caruana Galizia: anti-corruption activist, prominent investigative journalist.

India

Gauri Lankesh: human rights journalist, critic of caste system, and rightwing Hindu nationalism

Philippines

Leonela Tapdasan Pesadilla: farmer activist, opposed mining projects, donated land for indigenous school.

Mia Manuelita Mascariñas-Green: environmental land rights lawyer.

Guatemala

Laura Leonor Vásquez Pineda: community leader, opposing mining.

Honduras

Patricia Villamil Perdomo: defended migrant rights, spoke out against human trafficking groups, targeted by drug cartel

Argentina

Micaela Garcia: women's rights activist, campaigned against gender-based violence.

© Copyrights and Contacting Contributors

Copyrights for most of the work published in We'Moon belong to each individual contributor. Please honor the copyrights: ©: do not reproduce without the express permission of the artist, author, publisher or Mother Tongue Ink, depending on whose name follows the copyright sign. Some wemoon prefer to free the copyright on their work: ¤: this work may be passed on among women who wish to reprint it "in the spirit of We'Moon." In all cases, give credit to the author/artist and to We'Moon, and send each a copy. If the artist has given permission, We'Moon may release contact information. Contact mothertongue@wemoon.ws or contact contributors directly.

Contributor Bylines and Index
SEE PAGE 236 FOR INFO ABOUT HOW YOU CAN BECOME A WE'MOON CONTRIBUTOR!

Arna Baartz (Murwillumbah, Australia) Art is an expression of love. To see more please visit artofkundalini.com **p. 83**

Assetou Xango (Aurora, CO) is a poet and community activist born and raised in Colorado. Xango is dedicated to the rights of women and gender non-conforming people of color. To find out more, please search for her fanpage on Facebook. **p. 110, 141**

Autumn Skye ART (Powell River, BC) I offer my artwork as a mirror, as an intimate personal reflection and a grand archetypical revelation. Within these visions, may you recognize your own sacred heart, your cosmic divinity, and the innate grace that dwells within. autumnskyeart.com **p. 135, 145, 167, 177**

Barbara Dickinson (Sunny Valley, OR) I'm doing what I love, and loving what I do. Homesteading, creating, getting muddy, and communing. **p. 189**

Beate Metz (Berlin, Germany) was an astrologer, feminist, translator & mainstay of We'Moon's German edition & the European astrological community. **p. 205**

Beth Lenco (Hubbards, Nova Scotia) is a body/earth focused visual artist. She is the creator of Starflower Essences and a Shamanic healer. bethlenco.com **p. 128**

Bethroot Gwynn (Myrtle Creek, OR) 23 years as WeMoon's Special Editor & 43 at Fly Away Home women's land, growing food, theater & ritual. For info about spiritual gatherings, summertime visits send SASE to POB 593, Myrtle Creek, OR 97457. For info about her new poetry book, see p. 231. **p. 30, 31, 190**

Bettina 'Star-Rose' Madini (Montello, WI) I am so grateful to be on the planet here and now. What exciting times of change and new possibilities! I embrace all of it, every day! BettinaStar-Rose.com, bettinamadini@hotmail.com **p. 3**

Betty LaDuke (Ashland, OR) Honoring the earth and our common humanity. Through my art, publications, and work in the academic world, I honor and illuminate international women artists who have inspired my work. *Dreaming Cows* is a book depicting a 100-foot mural created in collaboration with Heifer International. Art photography by Robert Jaffe. bettyladuke.com **p. 23**

Carmen R. Sonnes (Phoenix, OR) creates paintings to bring beauty, healing and balance to our planet. By acknowledging culture, the feminine and spirit in each work, she fulfills her purpose. Visit carmenrsonnes.com; carmenrsonnes@yahoo.com **p. 4**

Carol Wylie (Saskatoon, SK) is a portrait artist living in Saskatoon on the Canadian Prairies and specializing in the joy of people and their faces. Carolwylie.ca **p. 160**

Catherine Molland (Santa Fe, NM) is a professional artist and organic farmer. Her recent series *Navigating the Unknown* tells the story of the hero's journey from the feminine perspective. Women warriors healing the planet through love for all species. Artwork: catherinemolland.com **p. 85**

Cathy Casper (Arvada, CO) lives on the rising sun side of the Rockies. She holds the vision of a way of living—where people appreciate the gifts of the Sun, Moon and all forms of life, including each other, with hearts open wide. **p. 93**

Cathy McClelland (Kings Beach, CA) paints from her heart and imagination. Her love for nature, cross-cultural mythical subjects, magical, sacred places and symbols fuel her creative spirit. cathymcclelland.com **p. 82**

Chasity Bleu (Keaau, HI) My greatest inspiration and passion is the Divine Feminine. I seek Her in all things. I am an artist and writer who loves yoga, crystals, Earl Grey, Nature, and the stars. chasitybleu.weebly.com   **p. 119**

Collin Stuart Chambers (Morehead City, NC) aka Madera Salvaje. Wild Woman Creatrix-Sacred Movement Artist-Ceremonial Healer Goddess-Maker of Magic Wands-Intuitive LightWorker of the Divine Feminine, and Awakener of Magic. WildWoodMagic.com   **p. 138**

Daisy Curley (Olympia, WA) Art continues to teach, heal, and inspire me. I'm finding myself in the process.   **p. 112**

Daughters of the Moon Tarot (Eugene, OR) Ffiona Morgan is an Elder Priestess/Lover of Goddess, feminist educator who teaches all aspects of the Craft. She visioned and painted *Daughters of the Moon Tarot*. A drummer, artist, writer and well known foremother of goddess spirit family. 541-654-0424, daughtersofthemoon.com   **p. 46, 159**

Deborah Koff-Chapin (Langley, WA) has been developing Touch Drawing since originating it in 1974. She is creator of *SoulCards 1&2, SoulTouch Coloring Journals* and author of *Drawing Out Your Soul*. Deborah teaches internationally and is an interpretive artist at numerous conferences. touchdrawing.com **p. 8**

Debra Hall (Castle Douglas, Scotland) I am a soulmaker, natural healer, mindfulness teacher and mentor. I like to be contacted about my art and soul creations at debra.ha@hotmail.co.uk and one-on-one Skype sessions. **p. 86, 102**

Denise Kester (Ashland, OR) is a mixed media printmaking artist, renowned teacher and founder/artistic director of Drawing on the Dream, an art distribution company. Announcing the new book *Drawing on the Dream— Finding My Way by Art*. drawingonthedream.com   **p. 87, 127**

Denise Ostler (Asheville, NC) writes, paints, doodles and plays ukulele in her garden cottage. You can read her stories and see her art on a Facebook page called FairyTale Medicine   **p. 47, 163**

Dhira Lawrence (Eugene, OR) opens a path of exploration with her beautiful healing images. Growing up surrounded by the natural beauty of Hawaii along with a spiritual upbringing, helped her form a powerful bond with Earth and all its many forms. dhiralawrence.com   **p. 180**

Diana Rivers (Elkins, AR) is an artist and writer living on women's land. She writes books about the Hadra—women with powers—as well as poetry and short stories. She has helped organize several venues for women in the arts. thehadra.com and rivers5524@aol.com   **p. 113**

Diane Norrie (Coquitlam, BC) Visual artist and tea cup reader. I live in the Fraser River Valley in Small Red Salmon, BC. I am strongly influenced by a spiritual connection; it makes my work constantly change and evolve.   **p. 146, 181**

Dorrie Joy (Somerset, UK) is an intuitive artist working in many mediums in celebration of our shared indigeny as people of Earth. Originals, prints, commissions of all kinds at dorriejoy.co.uk   **p. 43, 172**

Eileen M. Rose (Valley Stream, NY) co-author of *Create Your Own Sand Mandala for Meditation, Healing and Prayer*. A visionary artist, photographer, sacred geometer, sand mandala facilitator/drawing teacher; Her goal is to convey the divine, and beauty in nature, and to spread love—it is contagious. illuminatedrose.com, facebook, Instrgram at Illuminatedrose   **p. 39**

Elspeth McLean (Pender Island, Australia) uses colour, pattern and symbology in her artwork to create vibrant, uplifting pieces. She hopes her art connects the viewer with their inner child. "Express and celebrate the colours of your soul." elspethmclean.com **p. 185**

Emily Balivet (Pittsford, VT) Her artistic renderings of the divine feminine, mythological figures, ancient earth-based religions has won her a worldwide and multicultured fan base. She's inspired by the pre-Raphaelite/Art/Nouveau/1960's Psychedelic Art movements and the luscious landscapes of her native Vermont and Alaska. Her painting probes the nature of spirituality. EmilyBalivet. com **p. Front Cover, 4, 57, 178**

Emily Kedar (Toronto, ON) is a poet and therapist dedicated to nature worship and heart awakening. She lives and practices in Toronto and Salt Spring Island. **p 105, 143**

Emily Kell (Bulder, CO) A visionary artist, my work takes root in ideals of divine feminine and primordial goddess energy. I created a language to write secret poems and messages that I include in my female empowerment series honoring women in their many unique manifestations. emilykell.com, facebook at emilykellart **p. 45, 155**

Evangeline Coomber (Cornwall, UK) After studying Fine Art (sculpture) to degree level, I spent almost two decades in teaching. I now paint a variety of subjects in various media—from pastel portraits to mystical subjects in gouache-&-ink, and illustrations in watercolor. **p. 168**

Francene Hart (Honaunau, HI) is an internationally recognized visionary artist whose work utilizes the wisdom and symbolic imagery of Sacred Geometry, reverence for the natural environment and the interconnectedness between all things. francenehart.com **p. 17, 76**

Gael Nagle (Detroit, OR) has been batiking for 45 years and still finds it a challenging and inspiring medium. Gael has been painting as well lately. She and her husband, Charlie, live and work at Breitenbush Hot Springs. batiksbygael.com and etsy: batiksbygael **p. 4, back cover**

Gaia Orion (Sebright, ON) creates colorful geometric patterns, unveiling universal themes. By sharing with others, she discovered these transformational themes are part of a larger visionary movement: deep ecology, social justice, conscious politic, etc. gaiaorion.com **p. 36**

Gem Artemist (Edmonton, AB) aka Jen Aiken. My artistic creations and paper crafting are a quiet hobby that keeps me sane in a sometimes crazy world. mypaisleyheart.blogspot.com **p. 158**

Glenda Goodrich (Salem, OR) Mixed Media Artist and SoulCollage Facilitator. glendagoodrich.com **p. 169**

Gretchen Butler (Cazadero, CA) *Deep Time: An Art Odyssey*, my new book, explores evolutionary ramifications of the sun and earth's connections with stars. Scientific information inspires imagination. See more at gretchenbutlerwildartcafe.com **p. 77, 91**

Gretchen Lawlor (Seattle, WA and Tepaztlon, Mexico) We'Moon oracle, now mentor to new oracles & astrologers. Passionate about providing astrological perspective & support to worldwide wemoon re: creativity, work, love, $, health. My first book, *Windows of Time—Tools for Right Timing*, is available through Amazon. For consultations & info, contact me at light@whidbey.com; gretchenlawlor.com **p. 18**

Gwen Davies (Wales, UK) Her influences are folklore, nature and the Goddess. More of her work can be found at qwendavies.com and on Etsy at qwendaviesart **p. 27**

Heather Brunetti (Ashland, OR) Art is a whisper—a glimpse of the soul. It is a lasso cast out into the stars catching distant memories and bringing them back to earth. I celebrate sharing this journey with you. We are so fortunate. **p. 15**

Heather Rae Lawrence (Ashland, OR) is an artist from the Northwest; her artwork is inspired by visions she receives in dreams and mediations. Heather's work can be found at purpleheartist.com **p. 69, 92**

Heather Roan Robbins (Ronan, MT) ceremonialist, spiritual counselor, & astrologer for 40 years, author of *Moon Wisdom, Everyday Palmistry* & several children's books (Cico books, avail. on Amazon), writes weekly Starcodes column for We'Moon & The Santa Fe New Mexican, works by phone & Skype, practices in Montana, with working visits to Santa Fe, NM, MN, & NYC. roanrobbins.com **p. 12, 15, 228**

Hynden Walch (Los Angeles, CA) is an actress and urban pagan. **p. 97**

IAYAALIS Kali-Ma'at Eloai (Nashville, TN) is an Oracle, Shaman, Metaphysician, and certified Wholistic Life Coach; specializing in the Indigenous Healing Arts. She is the Creatrix & Co-Founder of Supernal-Indigenous Healing Arts Initiative. Iayaalis.com **p. 136**

Jakki Moore (Oslo, Norway) loves to travel & feels very grateful to use her art talents to bring attention & help those who need it most: animal, human & women's rights. jakkimoreart@yahoo.com & jakkiart.com **p. 123, 188**

Jan Kinney (Seattle, WA) drew inspiration from the *Tarot of Compassion* (yet unpublished) from the Pacific Northwest, where she lives. She also sings, preaches, and works with beads. Please send inquires about the *Tarot of Compassion* to janlkinney@gmail.com **p. 48, 131**

Jan Pellizzer (Grass Valley, CA) I am blessed to have found my creative spirit and nurture it with the inspiration around me. I live, love and create in the Sierra Foothills of Northern California. **p. 35**

Janet Newton (Peoria, IL) is a professor of graphic design at a community college in Illinois who enjoys drawing and painting tarot decks. **p. 103**

Janine Canan (Sonoma, CA) Poet, psychiatrist and volunteer for Amma. She lives in California's Valley of the Moon, and is the author of twenty-some books including: *Mystic Bliss, My Millennium: Culture, Spirituality & the Divine Feminine, Garland of Love, Ardor: Poems of Life* and *Journeys with Justine*. Visit JanineCanan.com & FB. **p. 161**

Janis Dyck (Golden, BC) works from the creative feminine fire in its many forms; trying to stay open to what needs expression. She is grateful & honoured to share what is born from this process. She lives with her family, dog & chickens in the beautiful Rocky Mountains. janisdyck@persona.ca **p. 50, 72**

Jenna Weston (Hawthorne, FL) poet, professional artist & workshop presenter. She authored the books: *Written on the Leaves: Haiku Poetry & Linocut Prints* & *Breathing Together: Poems of Transcendence*. jennaweston.com **p. 50**

Jennifer Lothrigel (Lafayette, CA) is an intuitive reader, healer, poet and photographer. jenniferlothrigel.com **p. 153**

Jennifer Smith (Vancouver, WA) of Visual Life Savers Art, creates beautiful whimsical oil paintings to bring happiness and connection to the world. email: jennifer@visuallifesaversart.com website: visuallifesaversart.com **p. 68**

Jenny Hahn (Kansas City, MO) captures the inward journey through bold, colorful expression. As cofounder of Creative Nectar Studio, she offers workshops across the country using painting as a tool for mindfulness and self-discovery. View more at jenspaintings.com **p. 89, 152**

Jo Jayson (Harrison, NY) is a spiritual artist, teacher, author, and has channeled the Sacred Feminine in her words for 8 years now. JoJayson.com **p. 49**

Joan Zehnder (Louisville, KY) Artist and Licensed Art Therapist. My art is created from the awareness that Life calls us to BE: to be present, to listen with full attentiveness, to learn, and to grow. Her voice is heard in the elements that make up our earth body and give us sustenance. She speaks to us in powerful symbols of spirit and soul. **p. 115, 176**

Joanne M. Clarkson (Olympia, WA) Joanne's poetry collection *The Fates* won the Bright Hill Press annual contest and was published in 2017. She loves reading Palms and Tarot. Her website is JoanneClarkson.com Daily Tarot available on FB at Joanne the Psychic. **p. 37, 55**

Joanne Rocky Delaplaine (Bethesda, MD) a mother of astonishment and daughter of night. She thanks the sun every morning at dawn. rockydelaplaine.com **p. 34**

Jodi Reeb (Minneapolis, MN) is a mixed-media, full-time artist and teacher for over 20 years working with encaustic, printmaking, collage, acrylic paint and sculpture—integrating community and nature themes. She loves to travel and integrate those experiences into her artwork. jodireeb.com. **p. 78**

Karen Russo (Elmira, OR) My sculpture is a tapestry of sensual form, texture and color. The warp is made of clay and colored pigments—Gifts from the Earth. The weft is the ancestral and personal narrative inspired by the Feminine and Nature. karenrusso.studio karenrusso333@gmail.com **p. 147**

Kay Kemp (Houston, TX) creates mixed media collage and painting to touch the heart and stir the soul. Her visionary art celebrates the sacred Feminine to inspire empowerment and harmony. Women WILL change the world! SpiritWorks4U.com and KayKemp.com **p. 28, 70**

Kim Crown (Chicago, IL) is an artist and renaissance woman with a raging rash of wanderlust. **p. 188**

Kimberly Webber (Taos, NM) Love is the antidote. Choose love. Experience more art: KimberlyWebber.com **p. 107**

Leah Marie Dorion (Prince Albert, SK) is an Indigenous artist from Prince Albert, Saskatechenwan, Canada. leahdorion.ca **p. 104, 143**

Leah Markman (Williams, OR) aka Cerulean Tango, has been reading Tarot and studying astrology in the beautiful Applegate Valley of Southern Oregon. She spends her time in the sunshine with her dog, horse and VW Bus. She writes poetry and practices mounted archery. Leahdmarkman@gmail.com **p.25**

Linda James (Seattle, WA) is an intuitive watercolor painter, educator, and flower essence practitioner creating her life in the wondrous environment of the Pacific Northwest. lindajamesart.com **p. 1**

Linda M. Evans (Poplar Branch, NC) Be true to yourself. Your inner voice won't let you down. **p. 183**

Lindsay Carron (Los Angeles, CA) Her work lies at the crossroads of indigenous voice and environmental activism. She illustrated 3 children's books recounting Tlingit oral tradition tales. Lindsay's work represents the potential for humanity to realize a healthy future for the planet. lindsaycarron.com instagram: @lacarron **p. 63, 111**

Lindy Kehoe (Gold Hill, OR) creates with the intention of painting portals of lucid dreaming, healing, and wonder. May all beings be free. Lindykehoe.com **p. 67, 142**

Lisa de St. Croix (Santa Fe, NM) is a metaphysical artist, creator of *Tarot De St. Croix* and *De St. Croix Lenormand*. Lisa teaches art and tarot workshops in her studio, globally and as workshops. Decks and info at lisadestcroix.com **p. 24**

Lisa J. Rough (Black Mountain, NC) is a self-taught, visionary writer and humble mama. Her ongoing work is to give the WILD a distinct voice. Visit her wild imaginings at LisaJRough.com **p. 61**

Lisa Kemmerer (Billings, MT) Ecofeminist, professor, and advocate for anymals, environment, and disempowered humans, Lisa has authored nine books: including *Eating Earth: Dietary Choice and Environmental Health*; *Speaking up for Animals: An Anthology of Women's Voices; Sister Species: Women, Animals, and Social Justice.* lisakemmerer.com **p. 66**

Lisette Costanzo (St. Catharines, ON) is an Intuitive Artist and Reiki Master. Her artwork focuses primarily on the Metaphysical as she utilizes high realism with otherworldly themes, inspired by the use of dreams, ritual, sound, music and nature. lisettecostanzo.com **p. 101**

Lissa Callirhoe (Albuquerque, NM) I coordinate a simultaneous singing to bless Mama Earth. We sing wherever we are, all singing the same song at 8:00 AM NM time, on Dark of the Moon and Full Moon, as per We'Moon calendar. Contact lissa@singingblessings.com **p. 88**

Liz Darling (Pittsburg, KS) is a visual artist and teacher. Deliberate and intricate, Darling creates precise compositions that center on themes of spirituality, transience, the divine feminine, and the natural world. lizdarlingart.com **p. 88**

Lorye Keats Hopper (Glastonbury, UK) I enjoy sharing creative empowerment through art. Writing, inner journey work and movement—and joining women's group circles in ceremony, to drum, dance and chant. **p. 38**

Lucy H. Pearce (Cork, Ireland) is a prolific author of transformational women's non-fiction: *Burning Woman; Moon Time; The Rainbow Way; Full Circle Health*. A passionate voice for women's empowerment & creativity, she is a vibrant artist & founder of WomanCraft Publishing. Lucyhpearce.com **p. 53, 114**

Lyndia Radice (Magdalena, NM) I live in rural New Mexico, and I paint the landscape and animals near my house. I am a photographer and musician. lyndiaradice.myportfolio.com **p. 156**

Mahada Thomas (Penticton, BC) Healer, writer, dreamer and soul singer. She shares her healing journey to inspire others. madathomas@yahoo.ca **p. 173**

Margarita Palma (Dunedin, FL) is a member of the Women's Mysteries at Enchanted Earth in Dunedin. Enchantedearthdunedin.com **p. 47**

Mari Susan Selby (Staunton, VA) has always been an unruly woman, author of *Lightning Strikes Twice*, 2X cancer survivor, publicist for authors who make a difference. selbyink.com mselby@gmail.com **p. 113, 146**

Marianna Jones (Milwaukie OR) is a poet, blogger, scholar and teacher. She makes sense of life through exploration of the written word. You can follow her work and connect with her at joyfilledheart.com **p. 157, 165**

Marleine Rose (Denver, CO) Marleine's poetry is predominantly channeled. She trusts the truth of the words that come through even when she doesn't fully understand them. She also receives messages in dreams which she writes as poetry. Contact her through Mystic Healing Muse and mrypoet@gmail.com **p. 43**

Marty Hamed (Manteo, NC) lives on Roanoke Island in the Outer Banks of North Carolina. She practices reflexology and Reiki, worships the Moon, runs with wolves and sometimes writes poetry. **p. 75**

McKenzie Brill (San Francisco, CA) Her passions include photography, travel and women's rights. She has long believed in writing as a powerful path to healing and connecting. Contact her at macbrill88@gmail.com or follow her on instagram at mckenzie.brill **p. 115**

Melanie Gendron (Felton, CA) "May you be showered with blessings"— Published artist, author, poet and creator of *The Gendron Tarot*. She manages Gendron Studios in the Santa Cruz mountains of California. melaniegendron.com **p. 33**

Melissa Renteria (Santa Cruz, CA) I love making art, surfing and the ocean. I have been collecting We'Moon calendar books for as long as I can remember. You can reach me about art at renterialissie@yahoo.com **p. 170**

Molly Brown (Bolinas, CA) is a painter, natural science illustrator, art teacher, shamanic practitioner, herbal potion maker and. . . lover of nature, dance and chocolate. mollybrownartist.com **p. 44**

MoonCat! (Montana, Texas, Florida) Traveling Astrologer, Artist, Radio DJ, Photographer, Jewelry Creator, PostCard Sender, GoddessCard Inventor, SeerofPatterns, Adventurer, Sagittarius. See CatOvertheMoon. com, LifeMapAstrology.com, Lifemapastrology@gmail.com **p. 204**

Mori Natura (Redwood Valley, CA) After a dozen years homesteading in the hills, I have become a gypsy nomad, writing from the road. I spend my days interacting with Mother Nature, caring for my son, and being in love. **p. 129**

Musawa (Estacada, OR & Tesuque, NM)I am hopeful that this year, I will be happily engaged in doing at least one of the many things I said I would be doing in last year's by-line—looking forward to becoming a liberated We'Moon Crone with more time for the creative life-affirming Gaia Revolution! **p. 6, 28, 186, 188**

Nancy Watterson (Oakland, OR) I am an artist, mother, grandmother, teacher, and gardener. In these ways I connect to the eternal and send my arrows into the future. Wherever I go, I find powerful women quietly nurturing and creating the becoming world. nwattersonscharf.com **p. 157**

Nell Aurelia (Devon, UK) lives with her daughters, connecting with land, finding a spiraling path from the ancestors to herself as a woman through words, walks, trees, dressing up for the soul and motherhood. nell.aurelia. admiral@gmail.com, thesingingdark.wordpress.com **p. 63, 69, 76, 117, 163**

Nicole Mizoguchi (Atlanta, GA) Her divinely inspired art reveals the mystical worlds of the soul, visions and dreams, She creates colorful images that radiate love, joy and healing energy to the hearts and minds of all who see her work. nicolemiz.com **p. 64**

Oak Chezar (Jamestown, CO) a radical dyke, performance artist, Women's Studies professor, psychotherapist, writer, & semi-retired barbarian. She lives in a straw bale, womyn-built house. She just published a memoir about Greenham Common Womyn's Peace Camp. Whilst working & playing towards the decimation of patriarchy & industrial civilization, she carries water. oakchezar@gmail.com **p. 98, 134**

The Obsidian Kat (Silver Springs, MD) Lucky number seven. The child who waited for the fairies, under the pear tree. A conscious curator of wellness. Mother and grandmother, all girls, divine feminine expressed. Happy crone. **p. 61**

OliviaJane Art (New Orleans, LA) I make art to look inside myself. I hope my art invites its viewers to look inside, to acknowledge our unique blend of masculinity and femininity. To heal the wars inside our body and mind, so we can work to heal the world around us. Instagram and FB at OliviaJaneArt and oliviajaneart.com **p. 25**

Pam Ballingham (Tucson, AZ) is a poet, visual artist and singer of the award-winning recording series, *Earth Mother Lullabies from Around the World.* Her poetry reflects the seamless interface between sound, and movement, image and language. earthmotherproductions.com **p. 56, 177**

Paula Franco (Buenos Aries, Argentina) Shaman woman, visual and visionary artist, teacher in sacred art, writer and poet, astrologer, tarot reader, creator of goddess cards and coloring book: *The Ancestral Goddess.* Paulafranco.net or pola_astroazul@hotmail.com **p. 73, 96**

Penn (Enid, OK) All is sacred under the Mother's sun. Blessings! **p. 121**

Raven Bishop (Church Hill, MD) is an artist & creative consultant. A ceremonialist artist working closely with the element of fire, she creates paintings & rituals that foster personal growth & community building. Raven is also a sculptor, graphic designer, performance & creative writer. ravendbishop.com **p. 41**

Rhea Wolf (Portland, OR) works to tear down the racist patriarchy through words, witchcraft, & workshops. The author of *The Light That Changes: The Moon in Astrology, Stories, and Time* and *Which is Witch* zines, Rhea is also a prison activist, mother, and teaches at the Portland School of Astrology. Keep in touch & visit RheaWolf.com **p. 19, 45, 59, 71, 81, 95, 107, 119, 131, 145, 155, 167, 179**

Rios de la Luz (El Paso, TX) is a queer xicana/chapina author of the short story collection, *The Pulse Between Dimensions* and *The Desert* (Ladybox Books) and the novella, *Itzá* (Broken River Books). She lives with the love of her life and her beautiful dog. riosdelaluz.wordpress.com **p. 65, 175**

Rita Loyd (Huntsville, AL) is a watercolor painter and writer. The theme of her work is about the healing power of unconditional self-love. Her art has appeared on over 100 magazine covers including *Science of Mind Magazine.* NurturingArt.com **p. 124, 183**

RM Allen (Exeter, NH) is an "artivist" who creates feminist and environmental art and writings with the goal of healing divides (ExeterNhArts.com). She is the author of the *New Hampshire Goddess Chronicles* series, and blogs at NhGoddess.com **p. 149**

Robin D. Bruce (Boulder, CO) is a healer, teacher, writer and singer living near the Flatirons of Boulder, Colorado. A current M.D.U student at Naropa University. Robin specializes in opening portals to spirit and wellness. You can find her at robindbruce.com **p. 126**

Robin Quinlivan (Thomas, WV) lives in the beautiful Appalachian Mountains of West Virginia, where she co-owns an art gallery. She is inspired by love for the natural world, as well as the mutable nature and interconnectedness of all things. robinquinlivan.etsy.com **p. 99**

Rose Flint (Wiltshire, England) has been a ceremonialist and Priestess-Poet for the Goddess Conference in Glastonbury. Her collection of Goddess poetry, *Grace, Breath, Bone* is available from roseflint9@gmail.com. Read more at www.poetrypf.co.uk/roseflint.shtml **p. 40, 109, 180**

Samantha Carney (Hanson, MA) I am a Massachusetts native, striving to nourish my coastal roots while navigating my nomadic heart. I am passionately writing, traveling, and loving my way through life as a devoted heart healer and community herbalist. doorwayoflight.com **p. 74**

Sandra Pastorius aka Laughing Giraffe (Ashland, OR) has been a practicing Astrologer since 1979, and writing for We'Moon since 1990. Look for her collected We'Moon essays under "Galactic Musings" at wemoon.ws. As a Gemini she delights in blending the playful and the profound. Email her about Birth chart readings and local Astrology Study Groups at: sandrapastorius@gmail.com. Peace Be! **p. 20, 22, 206**

Sandra Stanton (Farmington, ME) continues to explore the web of connections among people, other species & Mother Earth. More Goddesses are alive in *This Green World Oracle* with Kathleen Jenks, published by Schiffer. Goddessmyths.com or sandrastanton.com **p. 31**

Sandy Eastoak (Sebastopol, CA) is a shamanic painter and poet. She seeks climate and habitat healing through thanking plants and animals for their love and intelligence. Her art encourages celebrating all our relations. sandyeastoak.com **p. 168**

Sarah Cook (Danielsville, GA) is honored to be included in such an amazing group of creatix. Content to live simply in the woods with my man, animal and plant companions. Thankful for my drive to create art from the soul. **p. 164**

Serena Supplee (Moab, UT) has been "Artist on the Colorado Plateau" since 1980. An oil painter, watercolorist and sculptor, she welcomes you to visit her Moab studio by appointment or website serenasupplee.com **p. 84, 171, 230**

Shelley Anne Tipton Irish (Seattle, WA) Through passionate color and visionary storytelling, I infuse classically rendered oil paintings with transcendental exploration. It is my life ritual, sharing it with you, my bliss, best blessings! gallerysati.com **p. 105**

Shelley Blooms (Cleveland, OH) is a spirit pilgrim, ever on the trail of cosmic breadcrumbs by which the muses choose to amuse the noodle, through picture, poem, kit and caboodle. **p. 91**

Shoshanah Dubiner (Ashland, OR) is a psyche-morpho-grapher, a visionary artist who depicts forms best known to biologist—forms so varied & ingenious as to offer a language suitable for exploring the human psyche. cybermuse.com **p. 153**

Sisterdiscordia (Asheville, NC) is a writer and faery witch. Sharing her life with her true love and cat. When not working at Asheville Raven and Crone creating magic, she can be found barefoot in a creek or under a stack of books. In memory of Bonnie Frontino. Ladyshaper@hotmail.com **p. 160**

Sophia Rosenberg (Lasqueti Island, BC) lives in a small cabin in a big woods. sophiarosenberg.com or Bluebeetlestudio on Etsy **p. 62, 132, 149**

Stacie Haus (Spooner, WI) loves the process of intuitive painting and the excitement of following her heart and pushing her edge as the painting unfolds before her. **p. 60**

Stephanie A. Sellers (Fayetteville, PA) is a homesteader in the mountains. She owes healing to Grandmother Chickweed; and her favorite flowers include miniature jonquils, zinnia, and antique roses. She believes completely in the ability of women to heal themselves. sednasdaughters.com **p. 101, 124**

Sudie Rakusin (Hillsborough, NC) is a visual artist, sculptor, children's book author and illustrator for established authors Mary Daly and Carolyn Gage. She lives in the woods with her dogs, on the edge of a meadow, surrounded by her gardens. sudierakusin.com wingedwillowpress.com **p. 80, 116**

Sue Burns (Portland, OR) is a feminist, witch, writer, herbalist, teacher, mother. She is renewed in bodies of water, puts salt in her coffee, and looks for magick everywhere. **p. 4, 24**

Sue Davis (Fort Wayne, IN) My work is inspired by nature's spirit and the joy of exploration and discovery. I think about connectedness and reflect on a time when humanity held greater respect for the natural rhythms and cycles of the Earth and Moon. VachonArts.com **p. 174**

Sue Ellen Parkinson (Willits, CA) is an artist who firmly believes that imagery can support transformation. She paints the Sacred Feminine and lives in the mountains of Northern CA. Please contact her through her website: miracleofyourlife.com. She loves to connect with others! **p. 139**

Summer Rae (Brooksville, FL) Art is the reflection of self, for the viewer as much as the artist. I aim to inspire, evoke, and enlighten. Art for me, is a flower in a vast landscape of thorns. facebook.com/SummerRaeArt **p. 109**

Sundara Fawn (Marshall, NC) Yogini Spiritual Scientist, Artist, Author, Founder of Reawakening the Soul—A Journey to Discover and Express Your True Nature, empowerment program. Oracle card deck, online courses. SundaraFawn.com **p. 151**

Susa Silvermarie (Ajijic, Mexico) I turned 70 in 2017 and began a new life as an immigrant to Mexico. I blog, I write for several local publications and I run a weekly Write-to-a-prompt circle as well as a Writers Salon. I enjoy volunteering with an art program for Mexican children, and I truly love my life on the shore of Lake Chapala. Seeking local kindred spirits—come on down! susasilvermarie.com **p. 26, 49, 68, 83, 85, 104, 123, 142, 156, 159, 176**

Susan Baylies (Durham, NC) sells her lunar phases as cards, larger print charts and posters at snakeandsnake.com Email her at sbaylies@gmail.com **p. 226**

Susan Bolen (Mariposa, CA) is a reclusive painter who is intent on sending her art out into the world. She lives with her loving husband, 6 cats, and a chihuahua mix. Original art, prints and cards are available from Williams Gallery West, Oak Hurst, CA (559-693-5551); Manterbolen.com **p. 59**

Susan L. Roberts (Flushing, NY) studies Traditional Chinese Medicine, and other healing traditions, weaving them into her practice of occupational therapy and sustainable healthcare. Her VERY STRICT rules for life: 1) Have Fun! 2) Play Safely! 3) Make Friends! 4) Imagine! susanlroberts.com **p. 125**

Susan Levitt (San Francisco, CA) is a tarot card reader, astrologer, and feng shui consultant. She is the author of 5 books that are published in several languages including *Taoist Astrology* and *Introduction to Tarot*. Susan writes about new and full Moons on her blog at susanlevitt.com **p. 7, 23**

Susan Solari (Albuquerque, NM) is a visionary artist who studied painting and drawing at the Pennsylvania Academy of Fine Arts. She received her MFA and a Teaching Excellence Award at the University of Colorado. She is inspired by the spiritual and by nature. facebook.com/mysticalvisionsart ~ etsy.com/shop/MysticalVisionsArt ~ Mysticalvisions.net **p. 12**

Tamara Phillips (Vancouver, BC) art is inspired by the raw beauty of the natural world. Her watercolour paintings are woven together in earth tones, and she explores the connection between myth, dream, intuition, and reality. See more of her work here: TamaraPhillips.ca **p. 51, 71, 100, 150, 154**

Tarra "Lu" Louis-Charles (Charlotte, NC) is a self-taught, Haitian-American artist. Tarra's work explores the female form in the context of vulnerability, emotions, portraits, bold contrasts, Caribbean-inspired color palette, and abstract expressionism. You can see more of her art and crafts at tarraluart.com and on Etsy and Instagram. **p. 148**

Tessa Helweg-Larsen (Victoria, BC) I create my artwork to express my love and concern for my local bioregion. I strive to live an environmentally and socially sustainable lifestyle by living in community, growing food, practicing herbalism and appreciating the wild areas around me. tessahelweglarson@gmail.com apothecaryherbfarm.com **p. 95**

Toni Truesdale (Bidetorn, ME) Artist, muralist, teacher and illustrator, Toni celebrates women, the natural environment and the diversity of the world's cultures. Contact her at tonitruesdale@gmail.com. Prints and cards available through tonitruesdale.com **p. 133**

ERRORS/CORRECTIONS

In *We'Moon 2018* astro data, we mistakenly marked the points of extreme planetary closeness or distance from the Sun as ApG and PrG instead of ApH and PrH. Apologies for the misprint.

We appreciate all feedback that comes in, and continually strive to get closer to perfection. Please let us know if you find anything amiss, and check our website near the beginning of the year for any posted corrections for this edition of We'Moon.

KNOW YOURSELF—MAP OF PLANETARY INFLUENCES

Most people, when considering astrology's benefits, are familiar with their Sun Sign; however, each of the planets within our solar system has a specific part to play in the complete knowledge of "The Self." Here is a quick run-down of our planets' astrological effects:

⊙ **The Sun** represents our soul purpose. It is what we are here on Earth to do or accomplish, and it informs how we go about that task. It answers the age-old question "Why am I here?"

☽ **The Moon** represents our capacity to feel or empathize with those around us and within our own soul as well. It awakens our intuitive and emotional body.

☿ **Mercury** is "The Thinker," and involves our communication skills: what we say, our words, our voice, and our thoughts, including the Teacher/Student, Master/Apprentice mode. Mercury affects how we connect with all the media tools of the day—our computers, phones, and even the postal, publishing and recording systems!

♀ **Venus** is our recognition of love, art and beauty. Venus is harmony in its expressed form, as well as compassion, bliss and acceptance.

♂ **Mars** is our sense of "Get Up and GO!" It is the capacity to take action and do; it represents being in motion. It can also affect our temperament.

♃ **Jupiter** is our quest for truth, living the belief systems that we hold and walking the path of what those beliefs say about us. It involves an ever-expanding desire to educate the Self through knowledge toward higher law, the adventure and opportunity of being on that road—sometimes literally entailing travel and foreign or international culture, language and/or customs.

♄ **Saturn** is the task master: active when we set a goal or plan then work strongly and steadily toward achieving what we have set out to do. Saturn takes life seriously along the way and can be rather stern, putting on an extra load of responsibility and effort.

⚷ **Chiron** is the "Wounded Healer." It relates to what we have brought into this lifetime in order to learn how to fix it, to perfect it, make it the best that it can possibly be! This is where we compete with ourselves to better our own previous score. In addition, it connects to our health-body—physiological and nutritional.

♅ **Uranus** is our capacity to experience "The Revolution," freedom to do things our own way, exhibiting our individual expression or even "Going Rogue" as we blast towards a future collective vision. Uranus inspires individual inclination to "Let me be ME" and connect to an ocean of humanity doing the same.

♆ **Neptune** is the spiritual veil, our connection to our inner psychology and consciousness, leading to the experience of our soul. Psychic presence and mediumship are influenced here too.

♇ **Pluto** is transformation, death/rebirth energy—to the extreme. In order for the butterfly to emerge, the caterpillar that it once was must completely give up its life! No going back; burn the bridge down; the volcano of one's own power explodes. Stand upon the mountaintop and catch the lightning bolt in your hand!

Ascendant or Rising Sign: In addition, you must consider the sign of the zodiac that was on the horizon at the moment of your birth. Your Rising sign describes how we relate to the external world and how it relates back to us—what we look like, how others see us and how we see ourselves.

It is the combination of all of these elements that makes us unique among all other persons alive! We are like snowflakes in that way! Sharing a Sun sign is not enough to put you in a singular category. Here's to our greater understanding! Know Yourself!

MoonCat! © Mother Tongue Ink 2011

GODDESS PLANETS: CERES, PALLAS, JUNO AND VESTA

"Asteroids" are small planets, located between the inner, personal planets (Sun to Mars) that move more swiftly through the zodiac, and the outer, social and collective planets (Jupiter to Pluto) whose slower movements mark generational shifts. Ceres, Pallas, Juno and Vesta are faces of the Great Goddess who is reawakening in our consciousness now, quickening abilities so urgently needed to solve our many personal, social, ecological and political problems.

⚳ **Ceres** (Goddess of corn and harvest) symbolizes our ability to nourish ourselves and others in a substantial and metaphoric way. As in the Greek myth of Demeter and Persephone, she helps us to let go and die, to understand mother-daughter dynamics, to re-parent ourselves and to educate by our senses.

⚵ **Juno** (partner of Jupiter) shows us what kind of committed partnership we long for, our own individual way to find fulfillment in personal and professional partnering. She wants partners to be team-workers, with equal rights and responsibilities.

⚴ **Pallas** (Athena) is a symbol for our creative intelligence and often hints at the sacrifice of women's own creativity or the lack of respect for it. She brings to the fore father-daughter issues, and points to difficulties in linking head, heart and womb.

⚶ **Vesta** (Vestal Virgin/Fire Priestess) reminds us first and foremost that we belong to ourselves and are allowed to do so! She shows us how to regenerate, to activate our passion, and how to carefully watch over our inner fire in the storms of everyday life.

excerpt Beate Metz © Mother Tongue Ink 2009

Ephemeris 101

A Planetary Ephemeris provides astronomical data showing the daily positions of celestial bodies in our solar system.

The planets have individual and predictable orbits around the sun and pathways through the constellations that correlate with the astrological signs of the Zodiac. This regularity is useful for sky viewing and creating astro charts for a particular date.

The earliest astrologers used these ephemeris tables to calculate individual birth and event charts. These circular maps plot planetary positions and the aspects—angles of relationships—in a "state of the solar system" as a permanent representation of a moment in time. The ephemeris can then be consulted to find when real-time or "transiting" planets will be in the same sign and degree as planets in the birth or event chart. For instance, use the ephemerides to follow the Sun through the houses of your own birth chart, and journal on each day the Sun conjuncts a planet. The sun reveals or sheds light on a sign or house, allowing those qualities to shine and thrive. Ephemerides can also be used to look up dates of past events in your life to learn what planets were highlighted in your chart at that time. In addition, looking up dates for future plans can illuminate beneficial timing of available planetary energies. Contact Sandra Pastorius for complete exercises and more info.

Read across from a particular date, ephemerides provide the sign and degree of all the Planets, the Sun and Moon and nodes of the Moon on that day. The lower box on the page offers a quick look at Astro data such as, when a planet changes sign (an ingress occurs), outer planet aspects, and their change in direction or retrograde period and much more. The larger boxes represent two different months as labeled.) Use the Signs and Symbols at a Glance on page 229 to note the symbols or glyphs of planets, signs and aspects.

Sandra Pastorius © Mother Tongue Ink 2018

Moon: O hr=Midnight and Noon=12PM

Planet Glyphs (p. 229)

R=Planet Retrogrades shown in shaded boxes

Ingress: January 1st the Sun moves into 10° Capricorn

Day	Sid.Time	☉	0 hr ☽	Noon ☽	True ☊	☿
1 Tu	6 41 26	10♑15 21	12♏21 35	18♏49 44	26♋52.2	23♐51.2
2 W	6 45 22	11 16 31	25 14 11	1♐35 10	26R 50.0	25 19.2
3 Th	6 49 19	12 17 42	7♐52 52	14 07 28	26 47.6	26 47.7
4 F	6 53 15	13 18 52	20 19 08	26 28 04	26 45.5	28 16.7
5 Sa	6 57 12	14 20 03	2♑34 27	8♑38 27	26 43.9	29 46.2
6 Su	7 01 08	15 21 14	14 40 17	20 40 08	26 42.9	1♑16.2
7 M	7 05 05	16 22 25	26 38 13	2♒34 48	26D 42.6	2 46.7
8 Tu	7 09 01	17 23 36	8♒30 08	14 24 32	26 42.9	4 17.6

Mars Ingress Aries January 1 @ 2:21 PM

Astro Data	Dy Hr Mn	Planet Ingress	Dy Hr Mn	Last Aspe	
	Dy Hr Mn		Dy Hr Mn	Dy Hr Mn	
♂0N	2 0:56	♂ ♈	1 2:21	1 22:27	♀
♅ D	6 20:28	☿ ♑	5 3:41	4 17:43	☿
☊ D	7 0:06	♀ ♐	7 11:19	7 6:21	♀

2019 Asteroid Ephemeris

Ceres · Pallas · Juno · Vesta — Longitude (00:00 GMT)

2019	Ceres	Pallas	Juno	Vesta
JAN 1	20♏59.8	21♈31.7	20♉D36.8	14♒27.5
11	24 47.2	24 14.4	21 46.6	19 22.3
21	28 22.4	26 28.6	23 44.7	24 18.9
31	1♐43.5	28 09.6	26 23.4	29 16.4
FEB 10	4 47.8	29R28.9	29 ??	4♓41.3
20	7 32.3	29 32.3	3♊14.3	9 11.6
MAR 2	9 54.0	28 57.6	7 13.9	14 07.8
12	11 48.8	27 35.5	11 29.8	19 02.4
22	13 13.2	25 26.4	15 58.2	23 54.6
APR 1	14 03.3	22 40.9	20 35.6	28 43.9
11	14R15.7	19 36.2	25 19.9	3♈29.8
21	13 49.1	16 34.8	0♋08.8	8 11.5
MAY 1	12 44.3	13 57.1	5 00.7	12 48.5
11	11 06.1	11 58.3	9 54.4	17 20.2
21	9 04.1	10 46.5	14 48.7	21 45.5
31	6 51.2	10 22.6	19 42.7	26 04.0
JUN 10	4 42.3	10♈44.1	24 36.0	0♉14.4
20	2 51.6	11 46.0	29 27.7	4 15.4
30	1 29.4	13 22.5	4♌17.4	8 06.0
JUL 10	0 42.1	15 28.5	9 04.9	11 44.2
20	0♐31.8	17 59.3	13 49.6	15 07.9
30	0 57.3	20 50.6	18 31.3	18 14.8
AUG 9	1 56.2	23 59.4	23 09.6	21 01.7
19	3 24.9	27 22.4	27 44.1	23 25.2
29	5 19.4	0♉57.5	2♍14.6	25 21.2
SEP 8	7 36.6	4 42.7	6 40.5	26 44.7
18	10 12.9	8 36.2	11 01.3	27 31.1
28	13 05.5	12 36.6	15 16.5	27R36.0
OCT 8	16 12.1	16 42.6	19 25.3	26 56.4
18	19 30.4	20 51.0	23 27.6	25 33.2
28	22 58.5	25 00.6	27 20.0	23 31.9
NOV 7	26 35.0	29 23.1	1♎03.4	21 04.2
17	0♑18.2	3♊40.7	4 36.3	18 27.9
27	4 06.9	7 58.8	7 56.2	16 01.7
DEC 7	7 59.9	12 05.7	11 02.7	14 03.0
17	11 56.1	16 32.4	13 48.9	12 43.2
27	15 54.5	20 46.0	16 16.5	12 07.5
JAN 6	19♑54.1	24♊55.9	18♎20.7	12D16.4

Ceres · Pallas · Juno · Vesta — Declination

2019	Ceres	Pallas	Juno	Vesta
JAN 1	11S50.3	06S05.7	02S24.3	20S06.4
11	13 37.8	04 04.0	00N47.3	17 10.8
FEB 10	14 55.8	00 19.0	04 26.9	13 52.0
MAR 22	15 48.1	05N15.0	07 57.5	10 18.9
APR 11	16 21.6	12 00.4	10 56.9	06 40.6
MAY 1	16 45.2	18 20.3	13 22.0	00 05.6
21	17 07.0	22 31.0	14 36.5	00N18.3
JUN 10	18 01.1	23 24.4	14 52.6	03 24.1
30	18 42.1	21 30.0	13 52.3	08 16.9
JUL 20	19 39.9	18 54.3	12 35.5	09 54.4
AUG 9	20 54.0	16 01.2	10 11.2	10 55.0
29	22 17.7	13 06.5	07 47.4	11 17.5
SEP 18	23 42.1	10 21.6	05 13.0	11 03.0
OCT 8	24 57.6	07 55.5	02 36.7	10 16.9
28	25 56.5	06 55.6	00 00.7	08 13.6
NOV 17	26 32.4	06 20.6	02S05.5	06 13.2
DEC 7	26 41.3	03N41.4	03 53.3	08 02.4
27	26S31.5	03N34.9	05S04.5	08N52.9

Psyche · Eros · Lilith · Toro — Longitude

2019	Psyche	Eros	Lilith	Toro
JAN 1	2♒19.2	11♏R20.2	19♑53.3	6♈17.0
11	6 13.5	11D54.1	21 30.6	16 29.5
21	10 07.0	14 23.6	23 36.1	27 29.6
31	13 58.9	18 36.1	24 43.7	9♉16.8
FEB 10	17 48.0	23 13.6	25 17.2	21 47.4
20	21 33.1	0♐56.1	25R13.7	4♊52.2
MAR 2	25 13.0	8 21.6	24 32.5	18 18.9
12	28 46.4	16 24.4	23 14.8	1♋53.0
22	2♓11.6	24 21.5	21 26.4	15 18.1
APR 1	5 27.0	2♑30.5	19 16.9	28 18.7
11	8 30.6	10 31.5	16 59.0	10♌42.0
21	11 20.1	18 31.5	14 47.2	22 19.0
MAY 1	13 53.0	26 20.3	12 53.9	3♍06.4
11	16 06.1	3♒45.1	11 28.4	13 05.1
21	17 56.0	11 07.2	10 36.4	22 18.5
31	19 19.4	18 09.2	10 18.9	0♎51.8
JUN 10	20 12.2	24 59.5	10♑35.1	8 50.6
20	20 31.2	1♓41.4	11 22.3	16 19.8
30	20R14.3	8 13.4	12 37.1	23 24.3
JUL 10	19 21.2	14 37.3	14 16.0	0♏08.4
20	17 55.4	20 53.9	16 15.8	6 35.1
30	16 04.6	27 04.0	18 33.2	12 47.5
AUG 9	14 00.3	3♈09.0	21 05.8	18 48.1
19	11 57.4	9 06.1	23 51.1	24 38.7
29	10 10.1	15 06.1	26 47.3	0♐21.1
SEP 8	8 50.6	21 00.0	29 52.7	5 56.7
18	8 06.6	26 51.5	3♒05.7	11 26.4
28	8D01.4	2♉41.2	6 24.7	16 51.5
OCT 8	8 35.3	8 29.7	9 49.1	22 12.6
18	9 45.6	14 17.4	13 17.3	27 30.2
28	11 29.4	20 04.6	16 48.6	2♑45.2
NOV 7	13 42.5	25 51.9	20 21.9	7 57.7
17	16 21.1	1♊39.4	23 56.2	13 08.1
27	19 21.6	7 27.4	27 30.5	18 09.9
DEC 7	22 40.9	13 16.3	1♒03.8	23 09.6
17	26 16.1	19 06.2	4 34.9	28 01.5
27	0♈04.8	24 57.4	8 02.9	3♒33.9
JAN 6	4♈04.5	0♒50.1	11♒26.3	8♒37.0

Psyche · Eros · Lilith · Toro — Declination

2019	Psyche	Eros	Lilith	Toro
JAN 1	21S22.7	51N06.4	12S54.2	14S24.0
11	21 04.0	34 17.5	14 47.8	07 42.0
FEB 10	20 18.7	17 37.7	15 57.8	00N51.0
MAR 22	19 10.8	05 57.4	16 11.3	09 35.4
APR 11	17 46.8	01S04.6	15 51.9	16 09.5
MAY 1	16 15.3	05 24.1	15 32.1	18 57.7
21	14 47.3	08 28.4	15 09.6	18 57.8
JUN 10	12 55.4	13 42.7	14 50.2	15 25.0
30	12 59.8	16 20.1	13 34.9	10 44.8
JUL 20	13 53.7	18 55.4	11 40.8	05 36.9
AUG 9	15 21.6	21 19.9	12 28.1	01S33.0
29	16 49.2	23 24.1	13 11.3	05 52.9
SEP 18	17 46.3	24 58.0	15 05.6	10 02.8
OCT 8	18 00.6	25 52.1	17 03.2	13 58.7
28	17 31.5	25 59.0	18 56.4	17 36.3
NOV 17	16 21.6	25 13.3	20 39.0	20 49.5
DEC 7	14 34.0	23 32.8	22 05.8	23 32.6
27	12S12.7	20S58.6	23S12.8	25S39.2

Saffo · Amor · Pandora · Icarus — Longitude

2019	Saffo	Amor	Pandora	Icarus
JAN 1	13♈25.6	8♈16.2	27♈04.2	12♑01.4
11	12♈06.5	10 10.0	1♉33.4	21 36.0
21	11 39.6	12 28.6	6 09.8	21 36.0
31	12D02.7	15 08.1	10 52.1	26 ??
FEB 10	13 10.9	18 05.4	15 43.4	3♓10.5
20	14 57.2	21 17.8	20 40.1	20 30.1
MAR 2	17 16.2	24 43.0	25 41.0	25 23.7
12	20 00.7	28 19.3	0♊43.5	12 55.3
22	23 06.2	2♉05.3	5 45.0	15 23.7
APR 1	26 28.2	5 59.7	10 42.9	28 28.5
11	0♉08.3	10 00.7	15 33.9	12♈37.6
21	3 49.8	14 10.5	20 13.8	24 59.1
MAY 1	7 44.0	18 25.1	24 39.3	11♉45.3
11	11 44.9	22 46.2	28 47.1	24 24.0
21	15 50.0	1♊13.3	2♋33.5	5♊04.3
31	20 00.5	5 45.3	5 53.3	10 08.8
JUN 10	24 13.1	10 21.0	8 44.5	14 06.4
20	28 27.7	15 00.1	11 02.9	11 06.8
30	2♊43.6	19 42.0	12 46.4	16 34.8
JUL 10	7 00.2	24 26.3	13♋54.2	16R37.4
20	11 17.0	29 11.6	14R24.2	15 47.4
30	15 33.3	3♋58.9	14 15.9	12 31.3
AUG 9	19 48.9	8 46.0	13 27.8	8 46.2
19	24 03.1	13 33.3	12 10.4	15♋50.1
29	28 15.4	18 19.3	10 27.1	6 53.3
SEP 8	2♋25.3	23 05.3	8 25.3	17♌39.3
18	6 32.0	27 49.1	6 21.5	22 28.6
28	10 32.4	2♌30.1	4 23.6	26♍16.5
OCT 8	14 33.2	7 07.5	2 46.7	26♎18.8
18	18 25.7	11 40.2	1 40.2	24♏47.0
28	22 11.5	16 08.8	1D15.2	6♐08.2
NOV 7	25 49.4	20 32.0	1♋35.2	14♑55.0
17	29 16.3	24 49.3	2 40.3	25 47.0
27	2♌31.1	29 00.7	4 27.5	9♒44.9
DEC 7	5 33.2	3♍05.8	6 52.5	29 44.3
17	8 17.4	7 04.5	9 52.2	17♓26.3
27	10 41.3	10 56.0	13 23.1	29 45.3
JAN 6	12♋40.9	14♍40.2	17♋19.6	15♈30.3

Saffo · Amor · Pandora · Icarus — Declination

2019	Saffo	Amor	Pandora	Icarus
JAN 1	11N36.9	03S55.4	14S54.8	26S45.3
11	11 44.9	02 01.6	11 03.6	26 33.2
FEB 10	12 38.0	00N01.7	06 55.5	25 51.4
MAR 22	13 48.0	02 50.6	02 30.5	24 53.1
APR 11	14 52.4	05 26.4	01N44.7	23 20.0
MAY 1	15 51.8	08 12.8	05 18.8	19 29.6
21	15 48.0	12 51.6	09 18.5	10 19.7
JUN 10	14 41.6	13 41.8	12 47.9	24 00.5
30	13 16.1	15 18.3	15 32.9	29 05.8
JUL 20	11 23.2	16 54.3	17 47.9	20 41.5
AUG 9	09 03.2	18 20.9	20 40.5	29 10.8
29	06 24.2	19 42.4	23 46.6	05 25.6
SEP 18	03 31.1	20 44.5	26 28.2	02 31.5
OCT 8	00 30.0	21 27.1	28 53.0	05 43.2
28	02S33.3	21 51.5	30 17.7	32 01.0
NOV 17	05 28.5	22 03.7	29 13.9	19 24.7
DEC 7	08 18.1	13 29.6	34 00.2	21 44.9
27	10S28.8	13S03.0	33N41.7	24S04.0

Diana · Hidalgo · Urania · Chiron — Longitude

2019	Diana	Hidalgo	Urania	Chiron
JAN 1	10♈37.0	29♑16.0	3♉50.4	28♓08.0
11	13 03.9	26♑35.7	7 54.5	28 23.0
21	15 56.9	23 48.8	12 15.8	28 42.7
31	19 12.3	21 36.1	16 51.0	29 06.9
FEB 10	22 46.9	20 23.6	21 37.7	29 34.4
20	26 37.0	20D53.3	26 33.5	0♈15.2
MAR 2	0♉42.9	22 04.9	1♊36.3	0 38.1
12	5 00.1	24 09.3	6 42.6	1 02.7
22	9 27.8	26 59.6	11 57.7	1 48.1
APR 1	14 00.5	0♒29.7	17 13.6	2 23.5
11	18 36.2	4♒57.3	22 31.7	2 58.2
21	23 14.0	9 26.4	27 51.0	3 15.1
MAY 1	27 51.8	13 45.8	3♋09.0	3 49.8
11	2♊30.7	18 47.3	8 30.7	4 31.8
21	7 08.4	22 28.9	13 49.8	4 56.4
31	11 44.6	26 19.4	19 08.7	5 17.7
JUN 10	16 19.3	29 38.6	24 29.0	5 47.0
20	20 38.6	2♓19.8	29 47.2	5 57.0
30	24 54.8	4♓51.2	5♌00.7	5R56.4
JUL 10	29 04.8	5♓12.6	10 00.7	5 53.3
20	3♋07.9	5 09.3	15 06.9	5 53.3
30	7 02.1	4 35.8	20 09.7	5 37.2
AUG 9	10 47.3	3 04.5	25 08.3	5 14.8
19	14 21.1	2♓10.5	0♍01.0	4 57.3
29	17 42.1	1 29.5	4 51.0	4 37.3
SEP 8	20 48.7	1 05.5	9 35.5	4 23.7
18	23 39.5	1 18.3	14 09.5	4 11.4
28	26 12.6	2♓08.9	18 37.3	3 56.7
OCT 8	28 25.5	3 13.0	22 55.9	3 44.1
18	0♌16.0	4 38.5	27 02.8	3 31.6
28	1 40.0	6 14.2	0♎58.3	3 24.4
NOV 7	2 41.0	8 07.0	4 38.6	3 02.4
17	3 11.9	10 01.2	8 00.2	1 44.3
27	3R07.6	12 04.9	11 01.7	1 33.2
DEC 7	2 28.7	14 20.3	13 42.0	1D26.4
17	1 18.1	16 40.5	15 42.7	1 31.3
27	29♋41.6	17 01.6	17 14.1	1♈41.6
JAN 6	29♋41.6	25♓24.1	18♎05.6	1♈41.6

Diana · Hidalgo · Urania · Chiron — Declination

2019	Diana	Hidalgo	Urania	Chiron
JAN 1	11N45.6	78N44.4	03N34.4	02N16.3
11	13 23.4	80 43.6	06 41.9	02 26.0
FEB 10	15 37.1	80 48.6	13 34.1	03 05.6
MAR 22	18 12.5	75 52.6	16 49.0	03 32.3
APR 11	23 32.9	69 45.2	19 41.3	03 59.3
MAY 1	27 38.2	64 08.0	21 30.5	04 05.4
21	26 55.8	59 41.8	22 31.7	04 10.1
JUN 10	28 18.6	57 37.2	23 10.5	04 12.5
30	27 59.5	54 01.0	24 32.9	04 04.9
JUL 20	26 51.6	52 50.7	25 56.9	05 15.3
AUG 9	26 29.6	51 39.1	27 13.4	04 54.1
29	24 12.8	49 58.0	28 21.6	04 32.7
SEP 18	21 04.0	50 57.3	28 28.0	04 11.4
OCT 8	19 25.5	03 05.8	25 03.0	03 51.0
28	13 26.4	00 57.1	21 17.4	03 31.6
NOV 17	09 21.5	00 05.8	18 21.3	03 31.6
DEC 7	05 17.8	00S56.9	06 00.0	03 22.0
27	01N49.1	25S24.1	18♏05.6	03N18.2

2019 Planetary Ephemeris

LONGITUDE — January 2019

Day	Sid.Time	☉	0 hr ☽	Noon ☽	True ☊	☿	♀	♂	⚳	♃	♄	♅	♆	♇
1 Tu	6 41 26	10♑15 21	12♏21 35	18♏49 44	26♋52.2	23♐51.2	23♏29.6	29♓56.1	20♏59.9	11♐46.2	11♑22.6	28♈36.9	14♓04.8	20♑35.6
2 W	6 45 22	11 16 31	25 14 11	1♐35 10	26R 50.0	25 19.2	24 28.7	0♈36.4	21 23.1	11 58.7	11 29.7	28R 36.6	14 06.1	20 37.6
3 Th	6 49 19	12 17 42	7♐52 52	14 07 28	26 47.6	26 47.7	25 28.3	1 16.7	21 46.2	12 11.3	11 36.8	28 36.4	14 07.4	20 39.6
4 F	6 53 15	13 18 52	20 19 08	26 28 04	26 45.5	28 16.7	26 28.4	1 57.1	22 09.2	12 23.7	11 43.9	28 36.2	14 08.7	20 41.6
5 Sa	6 57 12	14 20 03	2♑34 27	8♑38 27	26 43.9	29 46.2	27 28.8	2 37.5	22 32.1	12 36.2	11 51.0	28 36.1	14 10.0	20 43.7
6 Su	7 01 08	15 21 14	14 40 17	20 40 08	26 42.9	1♑16.2	28 29.7	3 18.0	22 54.9	12 48.5	11 58.1	28D 36.0	14 11.4	20 45.7
7 M	7 05 05	16 22 25	26 38 13	2♒34 48	26D 42.6	2 46.7	29 31.0	3 58.4	23 17.6	13 00.9	12 05.1	28 36.0	14 12.8	20 47.7
8 Tu	7 09 01	17 23 36	8♒30 08	14 24 32	26 42.9	4 17.6	0♐32.6	4 38.9	23 40.2	13 13.1	12 12.2	28 36.0	14 14.2	20 49.7
9 W	7 12 58	18 24 46	20 18 17	26 11 47	26 43.5	5 48.9	1 34.7	5 19.4	24 02.6	13 25.3	12 19.3	28 36.1	14 15.6	20 51.8
10 Th	7 16 55	19 25 56	2♓05 23	7♓59 33	26 44.3	7 20.7	2 37.0	6 00.0	24 25.0	13 37.5	12 26.4	28 36.3	14 17.1	20 53.8
11 F	7 20 51	20 27 06	13 54 43	19 51 21	26 45.1	8 52.9	3 39.8	6 40.5	24 47.2	13 49.5	12 33.4	28 36.5	14 18.6	20 55.8
12 Sa	7 24 48	21 28 15	25 50 00	1♈51 11	26 45.8	10 25.6	4 42.8	7 21.1	25 09.3	14 01.6	12 40.5	28 36.7	14 20.1	20 57.8
13 Su	7 28 44	22 29 23	7♈55 28	14 03 23	26 46.2	11 58.7	5 46.2	8 01.6	25 31.3	14 13.5	12 47.5	28 37.0	14 21.7	20 59.9
14 M	7 32 41	23 30 31	20 15 31	26 32 25	26R 46.3	13 32.3	6 49.9	8 42.2	25 53.1	14 25.4	12 54.5	28 37.3	14 23.2	21 01.9
15 Tu	7 36 37	24 31 39	2♉54 37	9♉22 34	26 46.3	15 06.3	7 53.9	9 22.8	26 14.9	14 37.2	13 01.5	28 37.7	14 24.8	21 03.9
16 W	7 40 34	25 32 45	15 56 44	22 37 25	26 46.2	16 40.9	8 58.1	10 03.4	26 36.5	14 48.9	13 08.5	28 38.2	14 26.5	21 06.0
17 Th	7 44 30	26 33 51	29 24 55	6♊19 19	26D 46.2	18 15.9	10 02.7	10 44.1	26 57.9	15 00.6	13 15.5	28 38.7	14 28.1	21 08.0
18 F	7 48 27	27 34 56	13♊20 37	20 28 38	26 46.2	19 51.4	11 07.5	11 24.7	27 19.3	15 12.2	13 22.4	28 39.3	14 29.8	21 10.0
19 Sa	7 52 24	28 36 01	27 43 02	5♋03 17	26 46.3	21 27.4	12 12.6	12 05.3	27 40.5	15 23.7	13 29.4	28 39.9	14 31.5	21 12.0
20 Su	7 56 20	29 37 05	12♋28 41	19 58 21	26R 46.5	23 04.0	13 18.0	12 45.9	28 01.5	15 35.2	13 36.3	28 40.5	14 33.2	21 14.0
21 M	8 00 17	0♒38 08	27 31 16	5♌06 20	26 46.6	24 41.1	14 23.6	13 26.6	28 22.4	15 46.6	13 43.2	28 41.2	14 35.0	21 16.0
22 Tu	8 04 13	1 39 10	12♌42 19	20 18 02	26 46.5	26 18.8	15 29.4	14 07.2	28 43.2	15 57.9	13 50.1	28 42.0	14 36.7	21 18.1
23 W	8 08 10	2 40 12	27 52 14	5♍23 48	26 46.0	27 57.0	16 35.5	14 47.9	29 03.9	16 09.1	13 56.9	28 42.8	14 38.5	21 20.1
24 Th	8 12 06	3 41 13	12♍51 42	20 15 03	26 45.3	29 35.9	17 41.8	15 28.5	29 24.4	16 20.2	14 03.8	28 43.7	14 40.4	21 22.0
25 F	8 16 03	4 42 13	27 33 07	4♎45 21	26 44.4	1♒15.3	18 48.4	16 09.1	29 44.7	16 31.3	14 10.6	28 44.6	14 42.2	21 24.0
26 Sa	8 19 59	5 43 13	11♎51 22	18 50 57	26 43.6	2 55.4	19 55.1	16 49.8	0♐04.9	16 42.2	14 17.4	28 45.6	14 44.0	21 26.0
27 Su	8 23 56	6 44 12	25 44 03	2♏30 45	26D 42.9	4 36.1	21 02.1	17 30.4	0 24.9	16 53.1	14 24.1	28 46.6	14 45.9	21 28.0
28 M	8 27 53	7 45 11	9♏11 14	15 45 48	26 42.7	6 17.4	22 09.2	18 11.1	0 44.8	17 03.9	14 30.8	28 47.7	14 47.8	21 30.0
29 Tu	8 31 49	8 46 10	22 14 47	28 38 36	26 43.1	7 59.4	23 16.6	18 51.7	1 04.6	17 14.6	14 37.5	28 48.8	14 49.7	21 31.9
30 W	8 35 46	9 47 07	4♐57 41	11♐12 28	26 43.9	9 42.1	24 24.1	19 32.3	1 24.1	17 25.2	14 44.2	28 49.9	14 51.6	21 33.9
31 Th	8 39 42	10 48 04	17 23 26	23 31 01	26 45.2	11 25.4	25 31.8	20 13.0	1 43.5	17 35.7	14 50.8	28 51.2	14 53.6	21 35.8

LONGITUDE — February 2019

Day	Sid.Time	☉	0 hr ☽	Noon ☽	True ☊	☿	♀	♂	⚳	♃	♄	♅	♆	♇
1 F	8 43 39	11♒49 01	29♐35 40	5♑37 46	26♋46.6	13♒09.4	26♐39.7	20♈53.6	2♐02.8	17♐46.2	14♑57.4	28♈52.4	14♓55.6	21♑37.8
2 Sa	8 47 35	12 49 56	11♑37 44	17 35 55	26 47.8	14 54.0	27 47.7	21 34.3	2 21.8	17 56.5	15 04.0	28 53.7	14 57.6	21 39.7
3 Su	8 51 32	13 50 51	23 32 41	29 28 19	26R 48.4	16 39.3	28 55.9	22 14.9	2 40.7	18 06.7	15 10.5	28 55.1	14 59.6	21 41.6
4 M	8 55 28	14 51 44	5♒23 07	11♒17 23	26 48.2	18 25.1	0♑04.3	22 55.6	2 59.4	18 16.8	15 17.0	28 56.5	15 01.6	21 43.5
5 Tu	8 59 25	15 52 37	17 11 21	23 05 17	26 46.9	20 11.6	1 12.8	23 36.2	3 18.0	18 26.9	15 23.5	28 58.0	15 03.6	21 45.4
6 W	9 03 22	16 53 28	28 59 26	4♓54 03	26 44.6	21 58.6	2 21.4	24 16.8	3 36.3	18 36.8	15 29.9	28 59.5	15 05.7	21 47.3
7 Th	9 07 18	17 54 18	10♓49 22	16 45 40	26 41.4	23 46.1	3 30.2	24 57.4	3 54.5	18 46.6	15 36.3	29 01.0	15 07.7	21 49.1
8 F	9 11 15	18 55 06	22 43 13	28 42 20	26 37.6	25 34.1	4 39.1	25 38.1	4 12.5	18 56.3	15 42.6	29 02.6	15 09.8	21 51.0
9 Sa	9 15 11	19 55 54	4♈43 20	10♈46 33	26 33.6	27 22.3	5 48.1	26 18.7	4 30.2	19 05.9	15 48.9	29 04.3	15 11.9	21 52.8
10 Su	9 19 08	20 56 39	16 52 23	23 01 13	26 29.8	29 10.8	6 57.2	26 59.3	4 47.8	19 15.4	15 55.2	29 06.0	15 14.0	21 54.7
11 M	9 23 04	21 57 24	29 13 28	5♉29 33	26 26.9	0♓59.5	8 06.5	27 39.9	5 05.2	19 24.7	16 01.4	29 07.7	15 16.1	21 56.5
12 Tu	9 27 01	22 58 06	11♉49 57	18 15 05	26D 25.1	2 48.0	9 15.9	28 20.5	5 22.4	19 34.0	16 07.6	29 09.5	15 18.3	21 58.3
13 W	9 30 57	23 58 47	24 45 24	1♊21 17	26 24.5	4 36.4	10 25.4	29 01.1	5 39.4	19 43.1	16 13.7	29 11.3	15 20.4	22 00.1
14 Th	9 34 54	24 59 27	8♊03 08	14 51 13	26 25.2	6 24.3	11 35.0	29 41.6	5 56.2	19 52.2	16 19.7	29 13.2	15 22.6	22 01.8
15 F	9 38 51	26 00 04	21 45 47	28 46 54	26 26.6	8 11.5	12 44.7	0♉22.2	6 12.7	20 01.1	16 25.8	29 15.1	15 24.7	22 03.6
16 Sa	9 42 47	27 00 41	5♋54 34	13♋08 35	26 28.1	9 57.7	13 54.5	1 02.7	6 29.1	20 09.8	16 31.7	29 17.1	15 26.9	22 05.3
17 Su	9 46 44	28 01 15	20 28 36	27 54 04	26R 29.2	11 42.6	15 04.4	1 43.3	6 45.2	20 18.5	16 37.7	29 19.1	15 29.1	22 07.1
18 M	9 50 40	29 01 48	5♌24 14	12♌58 12	26 29.0	13 25.7	16 14.4	2 23.8	7 01.2	20 27.0	16 43.5	29 21.1	15 31.3	22 08.8
19 Tu	9 54 37	0♓02 18	20 34 51	28 12 58	26 27.2	15 06.7	17 24.5	3 04.3	7 16.9	20 35.5	16 49.4	29 23.2	15 33.5	22 10.5
20 W	9 58 33	1 02 48	5♍51 15	13♍28 21	26 23.8	16 45.0	18 34.7	3 44.7	7 32.3	20 43.7	16 55.1	29 25.3	15 35.7	22 12.1
21 Th	10 02 30	2 03 15	21 02 58	28 33 52	26 18.9	18 20.1	19 45.0	4 25.2	7 47.6	20 51.9	17 00.8	29 27.5	15 38.0	22 13.8
22 F	10 06 26	3 03 42	5♎59 57	13♎20 16	26 13.3	19 51.5	20 55.4	5 05.6	8 02.6	20 59.9	17 06.5	29 29.7	15 40.2	22 15.4
23 Sa	10 10 23	4 04 06	20 34 06	27 40 57	26 07.7	21 18.5	22 05.8	5 46.1	8 17.4	21 07.8	17 12.1	29 31.9	15 42.4	22 17.0
24 Su	10 14 20	5 04 30	4♏40 29	11♏32 37	26 02.9	22 40.5	23 16.4	6 26.5	8 31.9	21 15.6	17 17.6	29 34.2	15 44.7	22 18.6
25 M	10 18 16	6 04 52	18 17 24	24 55 04	25 59.6	23 56.9	24 27.0	7 06.9	8 46.2	21 23.2	17 23.1	29 36.5	15 46.9	22 20.2
26 Tu	10 22 13	7 05 12	1♐25 58	7♐50 34	25D 58.0	25 07.0	25 37.8	7 47.2	9 00.3	21 30.7	17 28.5	29 38.9	15 49.2	22 21.8
27 W	10 26 09	8 05 32	14 09 24	20 23 01	25 58.0	26 10.2	26 48.6	8 27.6	9 14.1	21 38.0	17 33.9	29 41.3	15 51.4	22 23.3
28 Th	10 30 06	9 05 50	26 32 03	2♑37 07	25 59.2	27 06.0	27 59.4	9 08.0	9 27.6	21 45.3	17 39.2	29 43.7	15 53.7	22 24.8

Astro Data	Dy Hr Mn
♂0N	2 0:58
♅D	6 20:28
☊D	7 0:06
♃⚼♅	9 21:34
☽0N	13 7:53
♃□♆	13 18:59
☊R	14 6:56
☊D	17 2:51
☊R	20 23:04
☽0S	26 0:07
☊D	27 21:10
♄⚹♆	31 14:16
☊R	3 6:38
☽0N	9 14:02
☊D	12 22:12

Planet Ingress	Dy Hr Mn
♂ ♈	1 2:21
☿ ♑	5 3:41
♀ ♐	7 11:19
☉ ♒	20 9:01
☿ ♒	24 5:50
⚳ ♐	25 18:09
♀ ♑	3 22:30
☿ ♓	10 10:52
♂ ♉	14 10:52
☉ ♓	18 23:05
☊ R17	9:45
☽0S22	9:19
☊ D26	11:33
☿0N27	6:28

Last Aspect Dy Hr Mn		☽ Ingress	Dy Hr Mn
1 22:27	♀ ☌	♐	2 9:00
4 17:43	☿ ☌	♑	4 18:56
7 6:21	♀ ⚹	♒	7 6:47
9 16:54	♅ ⚹	♓	9 19:45
11 14:26	☉ ⚹	♈	12 8:19
14 15:57	♅ ☌	♉	14 18:32
16 18:35	☉ △	♊	17 1:01
19 1:34	♅ ⚹	♋	19 3:45
21 1:51	♅ □	♌	21 3:56
23 1:21	♅ △	♍	23 3:23
24 13:52	♇ △	♎	25 4:04
27 5:22	♅ ☍	♏	27 7:32
28 22:40	♇ ⚹	♐	29 14:34

Last Aspect Dy Hr Mn		☽ Ingress	Dy Hr Mn
31 22:34	♅ △	♑	1 0:48
3 10:54	♅ □	♒	3 13:04
6 0:00	♅ ⚹	♓	6 2:03
7 22:15	♇ ⚹	♈	8 14:35
10 23:49	♅ ☌	♉	11 1:30
12 22:27	☉ □	♊	13 9:33
15 12:50	♅ ⚹	♋	15 14:04
17 14:18	♅ □	♌	17 15:22
19 13:52	♅ △	♍	19 14:48
21 1:53	♇ △	♎	21 14:18
23 15:12	♅ ☍	♏	23 15:57
25 12:15	♀ ⚹	♐	25 21:21
28 6:18	♅ △	♑	28 6:49

☽ Phases & Eclipses Dy Hr Mn	
6 1:29	● 15♑25
6 1:42:40	● P 0.715
14 6:47	☽ 23♈48
21 5:17	○ 0♌52
21 5:13	● T 1.195
27 21:12	☾ 7♏38
4 21:05	● 15♒45
12 22:27	☽ 23♉55
19 15:55	○ 0♍42
26 11:29	☾ 7♐34

Astro Data

1 January 2019
Julian Day # 43465
SVP 4♓59'55"
GC 27♐06.3 ⚴ 21♎31.7
Eris 23♈00.0R ⚵ 20♉36.6
⚷ 28♓08.0 ⚶ 14♒27.6
☽ Mean ☊ 27♋34.0

1 February 2019
Julian Day # 43496
SVP 4♓59'50"
GC 27♐06.4 ⚴ 28♎17.6
Eris 23♈02.0 ⚵ 26♉41.2
⚷ 29♓09.3 ⚶ 29♒46.2
☽ Mean ☊ 25♋55.5

*Giving the positions of planets daily at midnight, Greenwich Mean Time (0:00 UT)
Each planet's retrograde period is shaded gray.

2019 Planetary Ephemeris

March 2019 — LONGITUDE

Day	Sid.Time	☉	0 hr ☽	Noon ☽	True ☊	☿	♀	♂	⚳	♃	♄	♅	♆	♇
1 F	10 34 02	10♓06 06	8♑38 49	14♑37 47	26♋00.7	27♓53.7	29♑10.4	9♉48.3	9♐40.9	21♐52.3	17♑44.4	29♈46.2	15♓56.0	22♑26.3
2 Sa	10 37 59	11 06 21	20 34 34	26 29 45	26R 01.9	28 32.9	0♒21.4	10 28.6	9 54.0	21 59.2	17 49.6	29 48.7	15 58.2	22 27.8
3 Su	10 41 55	12 06 34	2♒23 48	8♒17 14	26 01.8	29 03.3	1 32.5	11 08.9	10 06.7	22 06.0	17 54.7	29 51.2	16 00.5	22 29.3
4 M	10 45 52	13 06 46	14 10 27	20 03 51	25 59.8	29 24.4	2 43.6	11 49.2	10 19.2	22 12.7	17 59.7	29 53.8	16 02.8	22 30.7
5 Tu	10 49 49	14 06 56	25 57 46	1♓52 30	25 55.6	29R 36.2	3 54.8	12 29.5	10 31.4	22 19.1	18 04.7	29 56.4	16 05.1	22 32.1
6 W	10 53 45	15 07 04	7♓48 18	13 45 26	25 49.0	29 38.7	5 06.1	13 09.8	10 43.4	22 25.5	18 09.6	29 59.1	16 07.3	22 33.5
7 Th	10 57 42	16 07 10	19 44 04	25 44 23	25 40.5	29 31.9	6 17.4	13 50.0	10 55.0	22 31.6	18 14.4	0♉01.7	16 09.6	22 34.9
8 F	11 01 38	17 07 14	1♈46 33	7♈50 42	25 30.6	29 16.2	7 28.7	14 30.2	11 06.4	22 37.6	18 19.2	0 04.4	16 11.9	22 36.2
9 Sa	11 05 35	18 07 17	13 57 00	20 05 36	25 20.2	28 52.0	8 40.1	15 10.4	11 17.4	22 43.5	18 23.9	0 07.2	16 14.2	22 37.5
10 Su	11 09 31	19 07 17	26 16 39	2♉30 22	25 10.3	28 20.0	9 51.6	15 50.6	11 28.2	22 49.2	18 28.5	0 09.9	16 16.4	22 38.8
11 M	11 13 28	20 07 16	8♉46 55	15 06 34	25 01.9	27 40.9	11 03.1	16 30.8	11 38.7	22 54.7	18 33.0	0 12.7	16 18.7	22 40.1
12 Tu	11 17 24	21 07 12	21 29 33	27 56 09	24 55.5	26 55.8	12 14.7	17 11.0	11 48.8	23 00.1	18 37.5	0 15.6	16 21.0	22 41.3
13 W	11 21 21	22 07 06	4♊26 40	11♊01 26	24 51.7	26 05.7	13 26.3	17 51.1	11 58.7	23 05.3	18 41.9	0 18.4	16 23.3	22 42.6
14 Th	11 25 17	23 06 58	17 40 43	24 24 52	24D 50.1	25 11.9	14 37.9	18 31.2	12 08.2	23 10.4	18 46.2	0 21.3	16 25.5	22 43.8
15 F	11 29 14	24 06 48	1♋14 06	8♋08 40	24 50.1	24 15.6	15 49.6	19 11.3	12 17.5	23 15.3	18 50.5	0 24.2	16 27.8	22 44.9
16 Sa	11 33 11	25 06 35	15 08 42	22 14 14	24R 50.8	23 18.2	17 01.3	19 51.4	12 26.4	23 20.0	18 54.6	0 27.2	16 30.1	22 46.1
17 Su	11 37 07	26 06 21	29 25 11	6♌41 19	24 51.0	22 20.9	18 13.1	20 31.4	12 35.0	23 24.6	18 58.7	0 30.1	16 32.3	22 47.2
18 M	11 41 04	27 06 04	14♌02 14	21 27 22	24 49.6	21 24.9	19 24.9	21 11.4	12 43.3	23 29.0	19 02.7	0 33.1	16 34.6	22 48.3
19 Tu	11 45 00	28 05 44	28 55 56	6♍27 02	24 45.7	20 31.3	20 36.7	21 51.4	12 51.3	23 33.2	19 06.7	0 36.1	16 36.8	22 49.4
20 W	11 48 57	29 05 23	13♍59 35	21 32 22	24 39.3	19 41.1	21 48.6	22 31.4	12 58.9	23 37.3	19 10.5	0 39.1	16 39.1	22 50.4
21 Th	11 52 53	0♈04 59	29 04 11	6♎33 44	24 30.5	18 55.1	23 00.5	23 11.4	13 06.2	23 41.1	19 14.3	0 42.2	16 41.3	22 51.4
22 F	11 56 50	1 04 33	13♎59 51	21 21 24	24 20.3	18 13.9	24 12.5	23 51.3	13 13.1	23 44.8	19 18.0	0 45.3	16 43.5	22 52.4
23 Sa	12 00 46	2 04 06	28 37 28	5♏47 15	24 09.9	17 38.0	25 24.5	24 31.2	13 19.7	23 48.4	19 21.6	0 48.4	16 45.7	22 53.4
24 Su	12 04 43	3 03 36	12♏50 14	19 46 01	24 00.6	17 07.8	26 36.5	25 11.1	13 26.0	23 51.7	19 25.1	0 51.5	16 47.9	22 54.3
25 M	12 08 40	4 03 05	26 34 28	3♐15 39	23 53.1	16 43.5	27 48.6	25 51.0	13 31.9	23 54.9	19 28.5	0 54.7	16 50.1	22 55.2
26 Tu	12 12 36	5 02 32	9♐49 44	16 17 05	23 48.1	16 25.3	29 00.7	26 30.8	13 37.5	23 57.9	19 31.9	0 57.8	16 52.3	22 56.1
27 W	12 16 33	6 01 57	22 38 10	28 53 31	23 45.4	16 13.0	0♓12.8	27 10.6	13 42.7	24 00.8	19 35.2	1 01.0	16 54.5	22 57.0
28 Th	12 20 29	7 01 21	5♑03 46	11♑09 35	23D 44.6	16D 06.7	1 25.0	27 50.4	13 47.5	24 03.4	19 38.3	1 04.2	16 56.7	22 57.8
29 F	12 24 26	8 00 42	17 11 38	23 10 37	23R 44.8	16 06.2	2 37.2	28 30.2	13 52.0	24 05.9	19 41.4	1 07.4	16 58.9	22 58.6
30 Sa	12 28 22	9 00 02	29 07 12	5♒02 05	23 44.8	16 11.4	3 49.4	29 10.0	13 56.1	24 08.2	19 44.4	1 10.7	17 01.0	22 59.4
31 Su	12 32 19	9 59 21	10♒55 53	16 49 12	23 43.5	16 22.0	5 01.7	29 49.7	13 59.9	24 10.3	19 47.4	1 13.9	17 03.2	23 00.1

April 2019 — LONGITUDE

Day	Sid.Time	☉	0 hr ☽	Noon ☽	True ☊	☿	♀	♂	⚳	♃	♄	♅	♆	♇
1 M	12 36 15	10♈58 37	22♒42 37	28♒36 39	23♋40.1	16♓37.8	6♓14.0	0♊29.4	14♐03.2	24♐12.2	19♑50.2	1♉17.2	17♓05.3	23♑00.8
2 Tu	12 40 12	11 57 51	4♓31 45	10♓28 19	23R 34.0	16 58.6	7 26.3	1 09.1	14 06.2	24 13.9	19 52.9	1 20.5	17 07.4	23 01.5
3 W	12 44 09	12 57 04	16 26 44	22 27 15	23 25.0	17 24.1	8 38.6	1 48.8	14 08.8	24 15.4	19 55.6	1 23.8	17 09.5	23 02.1
4 Th	12 48 05	13 56 14	28 30 08	4♈35 34	23 13.5	17 54.1	9 51.0	2 28.5	14 11.0	24 16.8	19 58.1	1 27.1	17 11.6	23 02.8
5 F	12 52 02	14 55 23	10♈43 39	16 54 30	23 00.3	18 28.3	11 03.3	3 08.1	14 12.8	24 18.0	20 00.6	1 30.4	17 13.7	23 03.4
6 Sa	12 55 58	15 54 29	23 08 08	29 24 36	22 46.4	19 06.6	12 15.7	3 47.7	14 14.3	24 18.9	20 02.9	1 33.8	17 15.7	23 03.9
7 Su	12 59 55	16 53 34	5♉43 53	12♉05 58	22 33.0	19 48.8	13 28.2	4 27.3	14 15.3	24 19.7	20 05.2	1 37.1	17 17.8	23 04.5
8 M	13 03 51	17 52 36	18 30 52	24 58 33	22 21.4	20 34.5	14 40.6	5 06.9	14 16.0	24 20.3	20 07.4	1 40.5	17 19.8	23 05.0
9 Tu	13 07 48	18 51 36	1♊29 04	8♊02 27	22 12.3	21 23.7	15 53.0	5 46.5	14R 16.3	24 20.7	20 09.5	1 43.9	17 21.8	23 05.5
10 W	13 11 44	19 50 34	14 38 46	21 18 07	22 06.3	22 16.1	17 05.5	6 26.0	14 16.1	24R 21.0	20 11.5	1 47.3	17 23.8	23 05.9
11 Th	13 15 41	20 49 30	28 00 39	4♋46 29	22 03.0	23 11.6	18 18.0	7 05.5	14 15.6	24 21.0	20 13.4	1 50.7	17 25.8	23 06.3
12 F	13 19 37	21 48 24	11♋35 46	18 28 39	22 01.9	24 10.0	19 30.5	7 45.0	14 14.7	24 20.9	20 15.2	1 54.1	17 27.8	23 06.7
13 Sa	13 23 34	22 47 15	25 25 15	2♌25 37	22 01.8	25 11.2	20 43.0	8 24.5	14 13.4	24 20.5	20 16.9	1 57.5	17 29.7	23 07.1
14 Su	13 27 31	23 46 04	9♌29 47	16 37 39	22 01.4	26 15.1	21 55.5	9 03.9	14 11.7	24 20.0	20 18.5	2 00.9	17 31.7	23 07.4
15 M	13 31 27	24 44 51	23 49 00	1♍03 34	21 59.4	27 21.6	23 08.1	9 43.3	14 09.7	24 19.3	20 20.0	2 04.3	17 33.6	23 07.7
16 Tu	13 35 24	25 43 35	8♍20 51	15 40 16	21 55.0	28 30.5	24 20.6	10 22.7	14 07.2	24 18.4	20 21.5	2 07.7	17 35.5	23 08.0
17 W	13 39 20	26 42 17	23 01 07	0♎22 33	21 47.8	29 41.7	25 33.2	11 02.1	14 04.3	24 17.3	20 22.8	2 11.2	17 37.4	23 08.2
18 Th	13 43 17	27 40 57	7♎43 38	15 03 23	21 38.2	0♈55.3	26 45.8	11 41.5	14 01.0	24 16.0	20 24.0	2 14.6	17 39.2	23 08.4
19 F	13 47 13	28 39 35	22 20 50	29 35 01	21 26.9	2 11.0	27 58.4	12 20.8	13 57.4	24 14.6	20 25.1	2 18.0	17 41.0	23 08.6
20 Sa	13 51 10	29 38 11	6♏45 03	13♏50 11	21 15.3	3 28.9	29 11.0	13 00.1	13 53.4	24 12.9	20 26.2	2 21.5	17 42.9	23 08.8
21 Su	13 55 06	0♉36 45	20 49 46	27 43 20	21 04.6	4 48.8	0♈23.6	13 39.3	13 49.0	24 11.1	20 27.1	2 24.9	17 44.7	23 08.9
22 M	13 59 03	1 35 17	4♐30 36	11♐11 26	20 55.7	6 10.8	1 36.3	14 18.6	13 44.2	24 09.1	20 27.9	2 28.4	17 46.4	23 09.0
23 Tu	14 03 00	2 33 48	17 45 51	24 14 01	20 49.5	7 34.7	2 48.9	14 57.8	13 39.0	24 06.9	20 28.7	2 31.8	17 48.2	23 09.1
24 W	14 06 56	3 32 17	0♑36 16	6♑52 59	20 45.8	9 00.6	4 01.6	15 37.1	13 33.4	24 04.5	20 29.3	2 35.2	17 49.9	23R 09.1
25 Th	14 10 53	4 30 44	13 04 40	19 11 55	20D 44.3	10 28.4	5 14.3	16 16.2	13 27.5	24 02.0	20 29.9	2 38.7	17 51.6	23 09.1
26 F	14 14 49	5 29 09	25 15 20	1♒15 34	20R 44.1	11 58.1	6 27.0	16 55.4	13 21.2	23 59.3	20 30.3	2 42.1	17 53.3	23 09.1
27 Sa	14 18 46	6 27 33	7♒13 19	13 09 15	20 44.3	13 29.6	7 39.7	17 34.6	13 14.5	23 56.4	20 30.7	2 45.6	17 55.0	23 09.0
28 Su	14 22 42	7 25 56	19 04 04	24 58 25	20 43.7	15 03.0	8 52.5	18 13.7	13 07.5	23 53.3	20 30.9	2 49.0	17 56.6	23 09.0
29 M	14 26 39	8 24 17	0♓52 57	6♓48 17	20 41.4	16 38.2	10 05.2	18 52.8	13 00.1	23 50.0	20 31.1	2 52.5	17 58.3	23 08.8
30 Tu	14 30 35	9 22 36	12 44 59	18 43 33	20 36.7	18 15.3	11 18.0	19 31.9	12 52.3	23 46.6	20R 31.1	2 55.9	17 59.9	23 08.7

Astro Data	Planet Ingress	Last Aspect	☽ Ingress	Last Aspect	☽ Ingress	☽ Phases & Eclipses
Dy Hr Mn	Dy Hr Mn	Dy Hr Mn	Dy Hr Mn	Dy Hr Mn	Dy Hr Mn	Dy Hr Mn
☊ R 2 11:04	♀ ♒ 1 16:46	2 18:48 ♅ □	♒ 2 19:07	1 3:03 ♃ ⚹	♓ 1 14:49	6 16:05 ● 15♓47
☿ R 5 18:20	♅ ♉ 6 8:28	5 8:06 ♅ ⚹	♓ 5 8:12	3 15:37 ♃ □	♈ 4 2:58	14 10:28 ☽ 23♊33
♃⚺♇ 7 16:34	☉ ♈ 20 22:00	7 19:09 ☿ ☌	♈ 7 20:29	6 2:16 ♃ △	♉ 6 13:07	21 1:44 ○ 0♎09
☽0N 8 19:14	♀ ♓ 26 19:44	9 17:15 ♃ △	♉ 10 7:11	8 8:30 ♇ △	♊ 8 21:16	28 4:11 ☾ 7♑12
☊ D 14 10:15	♂ ♊ 31 6:13	12 9:32 ☿ ⚹	♊ 12 15:49	10 17:28 ♃ ☍	♋ 11 3:32	
☊ R 16 16:24		14 12:32 ☿ □	♋ 14 21:50	12 23:34 ☿ △	♌ 13 7:51	5 8:52 ● 15♈17
☿0S 16 21:56	☿ ♈ 17 6:02	16 18:04 ☉ △	♌ 17 0:58	15 1:40 ☉ △	♍ 15 10:15	12 19:07 ☽ 22♋35
☉0N 20 21:59	☉ ♉ 20 8:56	18 15:20 ♃ △	♍ 19 1:42	17 4:30 ♀ ☍	♎ 17 11:23	19 11:13 ○ 29♎07
☽0S 21 19:54	♀ ♈ 20 16:12	20 15:23 ♃ □	♎ 21 1:29	19 11:13 ☉ ☍	♏ 19 12:42	26 22:19 ☾ 6♒23
☊ D 28 4:18		22 18:11 ♀ △	♏ 23 2:17	21 4:01 ♇ ⚹	♐ 21 16:00	
☿ D 28 14:00		25 2:25 ♀ □	♐ 25 6:07	23 11:45 ♃ ☌	♑ 23 22:51	
☊ R 29 13:10		27 2:38 ♃ ☌	♑ 27 14:09	25 19:49 ♇ ☌	♒ 26 9:28	
☽0N 5 1:20		30 0:06 ♂ △	♒ 30 1:47	28 9:45 ♃ ⚹	♓ 28 22:13	
⚳ R 9 4:36						
♃ R 10 17:02						

Astro Data

1 March 2019
Julian Day # 43524
SVP 4♓59'46"
GC 27♐06.4 ⚴ 29♎02.9R
Eris 23♈12.2 ⚵ 6♊49.1
⚷ 0♈34.8 ⚶ 13♓38.4
☽ Mean ☊ 24♋26.5

1 April 2019
Julian Day # 43555
SVP 4♓59'43"
GC 27♐06.5 ⚴ 22♎40.7R
Eris 23♈30.1 ⚵ 20♊35.7
⚷ 2♈23.5 ⚶ 28♓44.0
☽ Mean ☊ 22♋48.0

*Giving the positions of planets daily at midnight, Greenwich Mean Time (0:00 UT)
Each planet's retrograde period is shaded gray.

2019 Planetary Ephemeris

LONGITUDE — May 2019

Day	Sid.Time	☉	0 hr ☽	Noon☽	True☊	☿	♀	♂	⚳	♃	♄	♅	♆	♇
1 W	14 34 32	10♉20 54	24♓44 29	0♈48 10	20♋29.6	19♈54.1	12♈30.7	20♊11.0	12♐44.2	23♐43.0	20♑31.1	2♉59.3	18♓01.4	23♑08.5
2 Th	14 38 29	11 19 10	6♈54 57	13 05 06	20R 20.1	21 34.8	13 43.5	20 50.0	12R 35.8	23R 39.2	20R 30.9	3 02.7	18 03.0	23R 08.3
3 F	14 42 25	12 17 24	19 18 47	25 36 10	20 08.9	23 17.3	14 56.3	21 29.1	12 27.0	23 35.2	20 30.7	3 06.2	18 04.5	23 08.1
4 Sa	14 46 22	13 15 37	1♉57 17	8♉22 06	19 57.1	25 01.6	16 09.1	22 08.1	12 17.9	23 31.1	20 30.4	3 09.6	18 06.0	23 07.9
5 Su	14 50 18	14 13 48	14 50 34	21 22 34	19 45.6	26 47.8	17 21.9	22 47.1	12 08.5	23 26.9	20 29.9	3 13.0	18 07.5	23 07.6
6 M	14 54 15	15 11 57	27 57 55	4♊36 26	19 35.7	28 35.8	18 34.7	23 26.1	11 58.8	23 22.4	20 29.4	3 16.4	18 08.9	23 07.3
7 Tu	14 58 11	16 10 05	11♊17 55	18 02 10	19 28.0	0♉25.6	19 47.5	24 05.1	11 48.8	23 17.8	20 28.8	3 19.8	18 10.3	23 06.9
8 W	15 02 08	17 08 11	24 48 59	1♋38 12	19 23.0	2 17.3	21 00.4	24 44.0	11 38.5	23 13.1	20 28.0	3 23.1	18 11.7	23 06.6
9 Th	15 06 04	18 06 16	8♋29 40	15 23 15	19D 20.6	4 10.8	22 13.2	25 22.9	11 27.9	23 08.2	20 27.2	3 26.5	18 13.1	23 06.2
10 F	15 10 01	19 04 18	22 18 52	29 16 26	19 20.1	6 06.1	23 26.0	26 01.9	11 17.1	23 03.1	20 26.3	3 29.9	18 14.4	23 05.7
11 Sa	15 13 58	20 02 18	6♌15 51	13♌17 05	19 20.7	8 03.3	24 38.9	26 40.7	11 06.0	22 57.9	20 25.3	3 33.2	18 15.7	23 05.3
12 Su	15 17 54	21 00 17	20 20 02	27 24 35	19R 21.2	10 02.2	25 51.7	27 19.6	10 54.7	22 52.6	20 24.2	3 36.5	18 17.0	23 04.8
13 M	15 21 51	21 58 13	4♍30 34	11♍37 46	19 20.5	12 02.9	27 04.6	27 58.5	10 43.1	22 47.1	20 23.0	3 39.9	18 18.3	23 04.3
14 Tu	15 25 47	22 56 08	18 45 55	25 54 39	19 17.9	14 05.3	28 17.4	28 37.3	10 31.3	22 41.5	20 21.7	3 43.2	18 19.5	23 03.8
15 W	15 29 44	23 54 01	3♎03 33	10♎12 06	19 13.1	16 09.3	29 30.3	29 16.1	10 19.3	22 35.8	20 20.3	3 46.5	18 20.7	23 03.2
16 Th	15 33 40	24 51 52	17 19 45	24 25 56	19 06.3	18 14.8	0♉43.2	29 54.9	10 07.2	22 29.9	20 18.8	3 49.7	18 21.9	23 02.6
17 F	15 37 37	25 49 42	1♏30 00	8♏31 20	18 58.1	20 21.7	1 56.0	0♋33.6	9 54.8	22 23.9	20 17.2	3 53.0	18 23.0	23 02.0
18 Sa	15 41 33	26 47 30	15 29 23	22 23 36	18 49.6	22 29.9	3 08.9	1 12.4	9 42.3	22 17.8	20 15.6	3 56.2	18 24.1	23 01.4
19 Su	15 45 30	27 45 16	29 13 30	5♐58 44	18 41.7	24 39.3	4 21.8	1 51.1	9 29.7	22 11.5	20 13.8	3 59.5	18 25.2	23 00.7
20 M	15 49 27	28 43 01	12♐39 02	19 14 13	18 35.2	26 49.5	5 34.7	2 29.8	9 16.9	22 05.2	20 12.0	4 02.7	18 26.3	23 00.0
21 Tu	15 53 23	29 40 45	25 44 15	2♑09 11	18 30.8	29 00.5	6 47.6	3 08.5	9 03.9	21 58.7	20 10.1	4 05.9	18 27.3	22 59.3
22 W	15 57 20	0♊38 28	8♑29 12	14 44 32	18D 28.4	1♊11.9	8 00.6	3 47.2	8 50.9	21 52.2	20 08.1	4 09.1	18 28.3	22 58.6
23 Th	16 01 16	1 36 09	20 55 32	27 02 38	18 27.9	3 23.5	9 13.5	4 25.8	8 37.8	21 45.5	20 06.0	4 12.2	18 29.2	22 57.8
24 F	16 05 13	2 33 50	3♒06 19	9♒07 07	18 28.8	5 35.1	10 26.5	5 04.5	8 24.6	21 38.8	20 03.8	4 15.4	18 30.2	22 57.1
25 Sa	16 09 09	3 31 29	15 05 36	21 02 23	18 30.2	7 46.3	11 39.4	5 43.1	8 11.3	21 31.9	20 01.5	4 18.5	18 31.1	22 56.3
26 Su	16 13 06	4 29 07	26 58 05	2♓53 21	18R 31.3	9 56.9	12 52.4	6 21.7	7 57.9	21 25.0	19 59.2	4 21.6	18 32.0	22 55.4
27 M	16 17 02	5 26 44	8♓48 48	14 45 06	18 31.6	12 06.6	14 05.4	7 00.3	7 44.6	21 18.0	19 56.8	4 24.6	18 32.8	22 54.6
28 Tu	16 20 59	6 24 20	20 42 51	26 42 39	18 30.3	14 15.2	15 18.4	7 38.9	7 31.2	21 10.9	19 54.3	4 27.7	18 33.6	22 53.7
29 W	16 24 56	7 21 56	2♈45 02	8♈50 31	18 27.4	16 22.4	16 31.4	8 17.4	7 17.8	21 03.7	19 51.7	4 30.7	18 34.4	22 52.8
30 Th	16 28 52	8 19 30	14 59 35	21 12 36	18 22.9	18 28.0	17 44.4	8 56.0	7 04.4	20 56.4	19 49.0	4 33.7	18 35.1	22 51.9
31 F	16 32 49	9 17 04	27 29 54	3♉51 43	18 17.2	20 31.7	18 57.4	9 34.5	6 51.0	20 49.1	19 46.2	4 36.7	18 35.9	22 50.9

LONGITUDE — June 2019

Day	Sid.Time	☉	0 hr ☽	Noon☽	True☊	☿	♀	♂	⚳	♃	♄	♅	♆	♇
1 Sa	16 36 45	10♊14 36	10♉18 14	16♉49 31	18♋10.8	22♊33.5	20♉10.4	10♋13.1	6♐37.7	20♐41.8	19♑43.4	4♉39.7	18♓36.5	22♑50.0
2 Su	16 40 42	11 12 08	23 25 31	0♊06 10	18R 04.6	24 33.1	21 23.5	10 51.6	6R 24.4	20R 34.3	19R 40.5	4 42.6	18 37.2	22R 49.0
3 M	16 44 38	12 09 39	6♊51 13	13 40 27	17 59.2	26 30.5	22 36.5	11 30.1	6 11.2	20 26.9	19 37.5	4 45.5	18 37.8	22 48.0
4 Tu	16 48 35	13 07 09	20 33 29	27 29 56	17 55.3	28 25.6	23 49.6	12 08.6	5 58.1	20 19.4	19 34.5	4 48.4	18 38.4	22 47.0
5 W	16 52 31	14 04 38	4♋29 23	11♋31 22	17D 53.0	0♋18.2	25 02.7	12 47.0	5 45.1	20 11.8	19 31.4	4 51.3	18 39.0	22 45.9
6 Th	16 56 28	15 02 06	18 35 26	25 41 07	17 52.3	2 08.2	26 15.7	13 25.5	5 32.2	20 04.2	19 28.2	4 54.1	18 39.5	22 44.9
7 F	17 00 25	15 59 32	2♌47 59	9♌55 37	17 52.9	3 55.8	27 28.8	14 03.9	5 19.5	19 56.6	19 24.9	4 56.9	18 40.0	22 43.8
8 Sa	17 04 21	16 56 58	17 03 38	24 11 42	17 54.3	5 40.7	28 41.9	14 42.4	5 06.8	19 49.0	19 21.6	4 59.7	18 40.4	22 42.7
9 Su	17 08 18	17 54 22	1♍19 29	8♍26 42	17 55.6	7 23.0	29 55.0	15 20.8	4 54.4	19 41.4	19 18.2	5 02.4	18 40.9	22 41.5
10 M	17 12 14	18 51 46	15 33 04	22 38 20	17R 56.4	9 02.6	1♊08.1	15 59.2	4 42.1	19 33.7	19 14.8	5 05.1	18 41.3	22 40.4
11 Tu	17 16 11	19 49 08	29 42 16	6♎44 37	17 56.1	10 39.6	2 21.2	16 37.6	4 30.0	19 26.0	19 11.3	5 07.8	18 41.6	22 39.2
12 W	17 20 07	20 46 28	13♎45 08	20 43 36	17 54.6	12 13.8	3 34.4	17 15.9	4 18.1	19 18.4	19 07.7	5 10.5	18 41.9	22 38.1
13 Th	17 24 04	21 43 48	27 39 46	4♏33 22	17 51.9	13 45.4	4 47.5	17 54.3	4 06.4	19 10.7	19 04.1	5 13.1	18 42.2	22 36.9
14 F	17 28 00	22 41 07	11♏24 11	18 11 58	17 48.6	15 14.1	6 00.6	18 32.6	3 55.0	19 03.1	19 00.4	5 15.7	18 42.5	22 35.7
15 Sa	17 31 57	23 38 26	24 56 30	1♐37 35	17 45.0	16 40.2	7 13.8	19 11.0	3 43.8	18 55.5	18 56.7	5 18.2	18 42.7	22 34.5
16 Su	17 35 54	24 35 43	8♐15 03	14 48 45	17 41.7	18 03.4	8 27.0	19 49.3	3 32.8	18 47.9	18 52.9	5 20.8	18 42.9	22 33.2
17 M	17 39 50	25 33 00	21 18 36	27 44 33	17 39.1	19 23.7	9 40.1	20 27.6	3 22.1	18 40.3	18 49.1	5 23.3	18 43.1	22 32.0
18 Tu	17 43 47	26 30 16	4♑06 37	10♑24 51	17 37.5	20 41.2	10 53.3	21 05.9	3 11.6	18 32.8	18 45.2	5 25.7	18 43.2	22 30.7
19 W	17 47 43	27 27 31	16 39 24	22 50 25	17D 36.9	21 55.8	12 06.5	21 44.2	3 01.4	18 25.3	18 41.3	5 28.1	18 43.3	22 29.4
20 Th	17 51 40	28 24 47	28 58 09	5♒02 53	17 37.3	23 07.4	13 19.8	22 22.5	2 51.5	18 17.8	18 37.3	5 30.5	18 43.4	22 28.1
21 F	17 55 36	29 22 01	11♒04 59	17 04 48	17 38.3	24 16.0	14 33.0	23 00.7	2 41.8	18 10.4	18 33.3	5 32.9	18R 43.4	22 26.8
22 Sa	17 59 33	0♋19 16	23 02 47	28 59 25	17 39.7	25 21.4	15 46.3	23 39.0	2 32.5	18 03.1	18 29.3	5 35.2	18 43.4	22 25.5
23 Su	18 03 30	1 16 30	4♓55 12	10♓50 39	17 41.1	26 23.7	16 59.5	24 17.2	2 23.5	17 55.8	18 25.2	5 37.5	18 43.4	22 24.2
24 M	18 07 26	2 13 44	16 46 21	22 42 52	17 42.2	27 22.7	18 12.8	24 55.4	2 14.8	17 48.5	18 21.0	5 39.7	18 43.4	22 22.9
25 Tu	18 11 23	3 10 58	28 40 48	4♈40 42	17R 42.8	28 18.4	19 26.1	25 33.7	2 06.4	17 41.4	18 16.9	5 41.9	18 43.3	22 21.5
26 W	18 15 19	4 08 11	10♈43 10	16 48 46	17 42.8	29 10.6	20 39.4	26 11.9	1 58.3	17 34.3	18 12.7	5 44.1	18 43.1	22 20.1
27 Th	18 19 16	5 05 25	22 58 02	29 11 29	17 42.1	29 59.3	21 52.8	26 50.1	1 50.5	17 27.2	18 08.4	5 46.2	18 43.0	22 18.8
28 F	18 23 12	6 02 39	5♉29 33	11♉52 39	17 41.0	0♌44.4	23 06.1	27 28.3	1 43.1	17 20.3	18 04.2	5 48.3	18 42.8	22 17.4
29 Sa	18 27 09	6 59 53	18 21 06	24 55 08	17 39.7	1 25.6	24 19.5	28 06.5	1 36.0	17 13.4	17 59.9	5 50.4	18 42.6	22 16.0
30 Su	18 31 05	7 57 06	1♊34 54	8♊20 24	17 38.4	2 03.0	25 32.9	28 44.7	1 29.3	17 06.7	17 55.6	5 52.4	18 42.3	22 14.6

Astro Data

	Dy Hr Mn
♅∠♆	2 2:51
☽0N	2 9:07
♃⊻♇	9 10:31
☊D	9 18:49
☊R	12 1:08
☽0S	15 14:38
☊D	22 19:12
☊R	26 16:47
☽0N	29 18:06
☊D	5 22:45
♃⚼♅	6 23:20
☊R	10 5:54
☽0S	11 21:11
♃⊻♄	14 16:29
♃□♆	16 15:23

Planet Ingress

	Dy Hr Mn
☿ ♉	6 18:26
♀ ♉	15 9:47
♂ ♋	16 3:10
☉ ♊	21 8:00
☿ ♊	21 10:53
☿ ♋	4 20:06
♀ ♊	9 1:38
☉ ♋	21 15:55
☿ ♌	27 0:21

Last Aspect / ☽ Ingress

Last Aspect Dy Hr Mn		☽ Ingress Dy Hr Mn
30 21:58	♃ □	♈ 1 10:25
3 8:48	☿ ☌	♉ 3 20:19
5 15:11	♇ △	♊ 6 3:41
7 23:51	♂ ☌	♋ 8 9:08
10 2:07	♀ □	♌ 10 13:15
12 12:26	♂ ⚹	♍ 12 16:23
14 17:20	♂ □	♎ 14 18:52
16 9:39	♇ □	♏ 16 21:27
18 21:13	☉ ☍	♐ 19 1:22
20 17:06	♃ ☌	♑ 21 7:57
23 3:59	♇ ☌	♒ 23 17:50
25 12:52	♃ ⚹	♓ 26 6:09
28 4:22	♇ ⚹	♈ 28 18:33
30 15:09	♇ □	♉ 31 4:44

Last Aspect / ☽ Ingress

Last Aspect Dy Hr Mn		☽ Ingress Dy Hr Mn
1 22:54	♇ △	♊ 2 11:49
4 15:43	☿ ☌	♋ 4 16:18
6 14:11	♀ ⚹	♌ 6 19:17
8 21:24	♀ □	♍ 8 21:46
10 12:03	♇ △	♎ 11 0:30
12 15:16	♇ □	♏ 13 4:04
14 19:47	♇ ⚹	♐ 15 9:04
17 8:32	☉ ☍	♑ 17 16:14
19 11:20	☿ ☍	♒ 20 2:02
21 14:03	♃ ⚹	♓ 22 14:03
24 23:11	☿ △	♈ 25 2:39
27 7:52	♂ □	♉ 27 13:33
29 18:39	♂ ⚹	♊ 29 21:10

☽ Phases & Eclipses

Dy Hr Mn	
4 22:47	● 14♉11
12 1:13	☽ 21♌03
18 21:13	○ 27♏39
26 16:35	☾ 5♓09
3 10:03	● 12♊34
10 6:00	☽ 19♍06
17 8:32	○ 25♐53
25 9:48	☾ 3♈34

Astro Data

1 May 2019
Julian Day # 43585
SVP 4♓59'40"
GC 27♐06.6 ⚵ 13♎56.9R
Eris 23♈49.6 ⚴ 5♋00.8
⚷ 4♈02.8 ⚶ 12♈48.6
☽ Mean ☊ 21♋12.7

1 June 2019
Julian Day # 43616
SVP 4♓59'36"
GC 27♐06.6 ⚵ 10♎22.7
Eris 24♈07.1 ⚴ 20♋12.2
⚷ 5♈19.7 ⚶ 26♈29.5
☽ Mean ☊ 19♋34.2

*Giving the positions of planets daily at midnight, Greenwich Mean Time (0:00 UT)
Each planet's retrograde period is shaded gray.

2019 Planetary Ephemeris

July 2019

LONGITUDE

Day	Sid.Time	☉	0 hr ☽	Noon ☽	True ☊	☿	♀	♂	⚳	♃	♄	♅	♆	♇
1 M	18 35 02	8♋54 20	15♊11 35	22♊08 13	17♋37.4	2♌36.3	26♊46.3	29♋22.9	1♐23.0	17♐00.0	17♑51.3	5♉54.4	18♓42.0	22♑13.2
2 Tu	18 38 59	9 51 34	29 09 59	6♋16 27	17R 36.7	3 05.5	27 59.7	0♌01.0	1R 16.9	16R 53.5	17R 46.9	5 56.3	18R 41.7	22R 11.8
3 W	18 42 55	10 48 48	13♋27 03	20 41 08	17D 36.4	3 30.5	29 13.1	0 39.2	1 11.3	16 47.0	17 42.6	5 58.2	18 41.4	22 10.4
4 Th	18 46 52	11 46 01	27 58 00	5♌16 51	17 36.5	3 51.1	0♋26.6	1 17.4	1 06.0	16 40.6	17 38.2	6 00.1	18 41.0	22 09.0
5 F	18 50 48	12 43 15	12♌36 55	19 57 22	17 36.7	4 07.2	1 40.0	1 55.6	1 01.1	16 34.4	17 33.8	6 01.9	18 40.6	22 07.5
6 Sa	18 54 45	13 40 28	27 17 27	4♍36 26	17 37.1	4 18.8	2 53.5	2 33.7	0 56.5	16 28.3	17 29.4	6 03.7	18 40.1	22 06.1
7 Su	18 58 41	14 37 41	11♍53 38	19 08 28	17 37.4	4R 25.7	4 07.0	3 11.9	0 52.3	16 22.3	17 25.0	6 05.4	18 39.6	22 04.6
8 M	19 02 38	15 34 54	26 20 26	3♎29 07	17 37.6	4 27.9	5 20.5	3 50.0	0 48.5	16 16.4	17 20.5	6 07.1	18 39.1	22 03.2
9 Tu	19 06 34	16 32 06	10♎34 12	17 35 27	17R 37.7	4 25.4	6 34.0	4 28.1	0 45.1	16 10.7	17 16.1	6 08.7	18 38.6	22 01.7
10 W	19 10 31	17 29 19	24 32 42	1♏25 53	17D 37.7	4 18.2	7 47.6	5 06.2	0 42.0	16 05.1	17 11.7	6 10.3	18 38.0	22 00.3
11 Th	19 14 28	18 26 31	8♏14 58	14 59 57	17 37.7	4 06.4	9 01.1	5 44.4	0 39.3	15 59.6	17 07.3	6 11.9	18 37.4	21 58.8
12 F	19 18 24	19 23 43	21 40 56	28 17 58	17 37.9	3 50.0	10 14.7	6 22.5	0 37.0	15 54.3	17 02.8	6 13.4	18 36.8	21 57.4
13 Sa	19 22 21	20 20 55	4♐51 11	11♐20 42	17 38.1	3 29.3	11 28.2	7 00.6	0 35.1	15 49.1	16 58.4	6 14.8	18 36.2	21 55.9
14 Su	19 26 17	21 18 07	17 46 38	24 09 10	17 38.4	3 04.5	12 41.8	7 38.7	0 33.5	15 44.0	16 54.0	6 16.3	18 35.5	21 54.5
15 M	19 30 14	22 15 19	0♑28 25	6♑44 32	17 38.8	2 35.9	13 55.4	8 16.8	0 32.3	15 39.1	16 49.6	6 17.6	18 34.8	21 53.0
16 Tu	19 34 10	23 12 32	12 57 41	19 08 03	17R 39.0	2 03.8	15 09.1	8 54.9	0 31.4	15 34.4	16 45.3	6 19.0	18 34.0	21 51.6
17 W	19 38 07	24 09 44	25 15 47	1♒21 05	17 38.9	1 28.8	16 22.7	9 32.9	0D 31.0	15 29.8	16 40.9	6 20.3	18 33.3	21 50.1
18 Th	19 42 03	25 06 58	7♒24 09	13 25 14	17 38.5	0 51.3	17 36.4	10 11.0	0 30.9	15 25.4	16 36.5	6 21.5	18 32.5	21 48.6
19 F	19 46 00	26 04 11	19 24 33	25 22 25	17 37.7	0 11.9	18 50.1	10 49.1	0 31.1	15 21.1	16 32.2	6 22.7	18 31.6	21 47.2
20 Sa	19 49 57	27 01 25	1♓19 06	7♓14 58	17 36.6	29♋31.4	20 03.8	11 27.2	0 31.7	15 16.9	16 27.9	6 23.8	18 30.8	21 45.7
21 Su	19 53 53	27 58 40	13 10 21	19 05 40	17 35.2	28 50.2	21 17.5	12 05.2	0 32.7	15 13.0	16 23.6	6 24.9	18 29.9	21 44.3
22 M	19 57 50	28 55 55	25 01 21	0♈57 50	17 33.7	28 09.3	22 31.2	12 43.3	0 34.0	15 09.2	16 19.3	6 26.0	18 29.0	21 42.8
23 Tu	20 01 46	29 53 11	6♈55 38	12 55 14	17 32.4	27 29.2	23 45.0	13 21.4	0 35.7	15 05.6	16 15.1	6 27.0	18 28.1	21 41.4
24 W	20 05 43	0♌50 27	18 57 09	25 01 57	17 31.5	26 50.7	24 58.8	13 59.4	0 37.7	15 02.1	16 10.9	6 27.9	18 27.1	21 40.0
25 Th	20 09 39	1 47 45	1♉10 11	7♉22 21	17D 31.1	26 14.6	26 12.6	14 37.5	0 40.1	14 58.8	16 06.7	6 28.9	18 26.1	21 38.5
26 F	20 13 36	2 45 04	13 39 02	20 00 42	17 31.4	25 41.5	27 26.4	15 15.6	0 42.9	14 55.7	16 02.5	6 29.7	18 25.1	21 37.1
27 Sa	20 17 32	3 42 23	26 27 49	3♊00 48	17 32.3	25 12.0	28 40.2	15 53.7	0 46.0	14 52.7	15 58.4	6 30.5	18 24.1	21 35.7
28 Su	20 21 29	4 39 43	9♊39 59	16 25 35	17 33.5	24 46.8	29 54.1	16 31.7	0 49.4	14 50.0	15 54.3	6 31.3	18 23.0	21 34.3
29 M	20 25 26	5 37 05	23 17 45	0♋16 27	17 34.7	24 26.2	1♌08.0	17 09.8	0 53.1	14 47.4	15 50.3	6 32.0	18 21.9	21 32.8
30 Tu	20 29 22	6 34 27	7♋21 32	14 32 41	17R 35.7	24 10.8	2 21.9	17 47.9	0 57.3	14 45.0	15 46.3	6 32.7	18 20.8	21 31.5
31 W	20 33 19	7 31 51	21 49 25	29 11 04	17 35.8	24 01.0	3 35.8	18 26.0	1 01.7	14 42.7	15 42.3	6 33.3	18 19.7	21 30.1

August 2019

LONGITUDE

Day	Sid.Time	☉	0 hr ☽	Noon ☽	True ☊	☿	♀	♂	⚳	♃	♄	♅	♆	♇
1 Th	20 37 15	8♌29 15	6♌36 50	14♌05 46	17♋35.0	23♋56.9	4♌49.8	19♌04.0	1♐06.5	14♐40.7	15♑38.4	6♉33.9	18♓18.5	21♑28.7
2 F	20 41 12	9 26 40	21 36 48	29 08 48	17R 33.2	23D 59.0	6 03.7	19 42.1	1 11.6	14R 38.8	15R 34.6	6 34.4	18R 17.4	21R 27.3
3 Sa	20 45 08	10 24 06	6♍40 39	14♍11 11	17 30.6	24 07.3	7 17.7	20 20.2	1 17.0	14 37.1	15 30.7	6 34.9	18 16.2	21 25.9
4 Su	20 49 05	11 21 32	21 39 20	29 04 09	17 27.5	24 22.1	8 31.7	20 58.3	1 22.7	14 35.6	15 27.0	6 35.3	18 14.9	21 24.6
5 M	20 53 01	12 18 59	6♎24 48	13♎40 35	17 24.5	24 43.3	9 45.7	21 36.4	1 28.8	14 34.3	15 23.3	6 35.7	18 13.7	21 23.2
6 Tu	20 56 58	13 16 27	20 51 01	27 55 44	17 22.1	25 11.0	10 59.7	22 14.4	1 35.2	14 33.2	15 19.6	6 36.0	18 12.4	21 21.9
7 W	21 00 55	14 13 55	4♏54 34	11♏47 27	17D 20.7	25 45.3	12 13.7	22 52.5	1 41.9	14 32.2	15 16.0	6 36.3	18 11.1	21 20.6
8 Th	21 04 51	15 11 25	18 34 29	25 15 50	17 20.4	26 26.0	13 27.8	23 30.6	1 48.9	14 31.5	15 12.5	6 36.5	18 09.8	21 19.3
9 F	21 08 48	16 08 55	1♐51 46	8♐22 36	17 21.2	27 13.1	14 41.9	24 08.7	1 56.2	14 30.9	15 09.0	6 36.7	18 08.5	21 18.0
10 Sa	21 12 44	17 06 26	14 48 42	21 10 27	17 22.7	28 06.6	15 55.9	24 46.8	2 03.8	14 30.5	15 05.6	6 36.8	18 07.2	21 16.7
11 Su	21 16 41	18 03 57	27 28 15	3♑42 28	17 24.3	29 06.3	17 10.0	25 24.9	2 11.7	14D 30.3	15 02.3	6 36.9	18 05.8	21 15.4
12 M	21 20 37	19 01 30	9♑53 30	16 01 43	17R 25.4	0♌12.0	18 24.1	26 03.0	2 19.8	14 30.3	14 59.0	6R 36.9	18 04.4	21 14.2
13 Tu	21 24 34	19 59 03	22 07 27	28 11 00	17 25.6	1 23.5	19 38.3	26 41.0	2 28.3	14 30.5	14 55.8	6 36.9	18 03.0	21 12.9
14 W	21 28 30	20 56 38	4♒12 42	10♒12 47	17 24.3	2 40.7	20 52.4	27 19.1	2 37.1	14 30.8	14 52.6	6 36.8	18 01.6	21 11.7
15 Th	21 32 27	21 54 14	16 11 32	22 09 10	17 21.3	4 03.2	22 06.5	27 57.2	2 46.1	14 31.4	14 49.5	6 36.7	18 00.2	21 10.5
16 F	21 36 24	22 51 50	28 05 55	4♓02 01	17 16.7	5 30.9	23 20.7	28 35.3	2 55.4	14 32.1	14 46.5	6 36.5	17 58.8	21 09.3
17 Sa	21 40 20	23 49 28	9♓57 42	15 53 10	17 10.7	7 03.3	24 34.9	29 13.4	3 04.9	14 33.0	14 43.6	6 36.3	17 57.3	21 08.1
18 Su	21 44 17	24 47 08	21 48 42	27 44 32	17 04.0	8 40.1	25 49.1	29 51.5	3 14.8	14 34.1	14 40.7	6 36.1	17 55.8	21 07.0
19 M	21 48 13	25 44 49	3♈40 58	9♈38 18	16 57.0	10 20.9	27 03.3	0♍29.7	3 24.9	14 35.4	14 38.0	6 35.7	17 54.3	21 05.8
20 Tu	21 52 10	26 42 31	15 36 54	21 37 08	16 50.6	12 05.4	28 17.5	1 07.8	3 35.2	14 36.8	14 35.2	6 35.4	17 52.8	21 04.7
21 W	21 56 06	27 40 15	27 39 25	3♉44 10	16 45.4	13 53.1	29 31.8	1 45.9	3 45.8	14 38.5	14 32.6	6 35.0	17 51.3	21 03.6
22 Th	22 00 03	28 38 00	9♉51 52	16 03 00	16 41.8	15 43.6	0♍46.0	2 24.0	3 56.7	14 40.3	14 30.1	6 34.5	17 49.8	21 02.5
23 F	22 03 59	29 35 47	22 18 05	28 37 38	16D 39.9	17 36.4	2 00.3	3 02.2	4 07.8	14 42.3	14 27.6	6 34.0	17 48.2	21 01.4
24 Sa	22 07 56	0♍33 36	5♊02 09	11♊32 09	16 39.7	19 31.2	3 14.6	3 40.3	4 19.1	14 44.5	14 25.2	6 33.4	17 46.7	21 00.4
25 Su	22 11 53	1 31 27	18 08 04	24 50 19	16 40.7	21 27.6	4 28.9	4 18.5	4 30.7	14 46.9	14 22.9	6 32.8	17 45.1	20 59.3
26 M	22 15 49	2 29 19	1♋39 13	8♋34 58	16 41.9	23 25.1	5 43.2	4 56.7	4 42.5	14 49.4	14 20.7	6 32.2	17 43.5	20 58.3
27 Tu	22 19 46	3 27 14	15 37 39	22 47 12	16R 42.6	25 23.4	6 57.6	5 34.9	4 54.6	14 52.1	14 18.5	6 31.5	17 42.0	20 57.3
28 W	22 23 42	4 25 10	0♌03 19	7♌25 32	16 41.9	27 22.2	8 11.9	6 13.0	5 06.9	14 55.0	14 16.5	6 30.7	17 40.4	20 56.4
29 Th	22 27 39	5 23 07	14 53 11	22 25 22	16 39.2	29 21.2	9 26.3	6 51.2	5 19.4	14 58.1	14 14.5	6 29.9	17 38.8	20 55.4
30 F	22 31 35	6 21 07	0♍01 00	7♍38 51	16 34.3	1♍20.2	10 40.7	7 29.4	5 32.2	15 01.4	14 12.6	6 29.1	17 37.2	20 54.5
31 Sa	22 35 32	7 19 07	15 17 34	22 55 46	16 27.7	3 18.9	11 55.1	8 07.7	5 45.2	15 04.8	14 10.8	6 28.2	17 35.6	20 53.6

Astro Data	Dy Hr Mn
☊ D	3 6:43
☿ R	7 23:16
☽ 0S	9 2:54
☊ R	9 11:42
☊ D	10 1:14
☊ R	16 9:15
⚳ D	17 19:07
☽ 0N	23 10:38
☊ D	25 1:23
☊ R	30 17:07
☿ D	1 3:59
☽ 0S	5 9:28
☊ D	7 17:18
♃ D	11 13:38
♅ R	12 2:28

Planet Ingress	Dy Hr Mn
♂ ♌	1 23:20
♀ ♋	3 15:19
☿ ♋R	19 7:07
☉ ♌	23 2:52
♀ ♌	28 1:55
☿ ♌	11 19:47
♂ ♍	18 5:19
♀ ♍	21 9:08
☉ ♍	23 10:03
☿ ♍	29 7:49

Last Aspect Dy Hr Mn		☽ Ingress	Dy Hr Mn
1 21:49	♀ ☌	♋	2 1:25
3 14:26	♇ ☍	♌	4 3:20
5 6:26	♃ △	♍	6 4:26
7 16:51	♇ △	♎	8 6:08
9 19:37	♇ □	♏	10 9:30
12 0:30	♇ ⚹	♐	12 15:06
14 1:31	♆ □	♑	14 23:06
16 21:39	☉ ☍	♒	17 9:20
18 15:55	♃ ⚹	♓	19 21:20
22 8:35	☉ △	♈	22 10:03
24 14:49	☿ □	♉	24 21:43
27 4:29	♀ ⚹	♊	27 6:30
28 15:25	♆ □	♋	29 11:32
31 3:34	☿ ☌	♌	31 13:19

Last Aspect Dy Hr Mn		☽ Ingress	Dy Hr Mn
1 20:49	♂ ☌	♍	2 13:22
4 4:28	☿ ⚹	♎	4 13:31
6 7:37	☿ □	♏	6 15:33
8 14:59	☿ △	♐	8 20:36
10 19:52	♂ △	♑	11 4:51
12 22:13	♇ ☌	♒	13 15:37
16 1:03	♂ ☍	♓	16 3:51
17 22:36	♇ ⚹	♈	18 16:34
21 4:08	♀ △	♉	21 4:38
22 21:34	♇ △	♊	23 14:35
25 7:00	☿ ⚹	♋	25 21:06
27 8:56	♇ ☍	♌	27 23:55
29 0:08	♃ △	♍	29 23:58
31 8:47	♇ △	♎	31 23:09

☽ Phases & Eclipses Dy Hr Mn	
2 19:17	● 10♋38
2 19:24:09	● T 04'33"
9 10:56	☽ 16♎58
16 21:39	○ 24♑04
16 21:32	● P 0.653
25 1:19	☾ 1♉51
1 3:13	● 8♌37
7 17:32	☽ 14♏56
15 12:30	○ 22♒24
23 14:57	☾ 0♊12
30 10:38	● 6♍47

Astro Data

1 July 2019
Julian Day # 43646
SVP 4♓59'31"
GC 27♐06.7 ⚵ 13♎33.8
Eris 24♈17.6 ⚴ 4♌46.3
⚷ 5♈54.7 ⚶ 8♉28.5
☽ Mean ☊ 17♋58.9

1 August 2019
Julian Day # 43677
SVP 4♓59'26"
GC 27♐06.8 ⚵ 21♎27.1
Eris 24♈19.1R ⚴ 19♌27.3
⚷ 5♈42.9R ⚶ 18♉49.9
☽ Mean ☊ 16♋20.4

*Giving the positions of planets daily at midnight, Greenwich Mean Time (0:00 UT)
Each planet's retrograde period is shaded gray.

2019 Planetary Ephemeris

LONGITUDE — September 2019

Day	Sid.Time	☉	0 hr ☽	Noon ☽	True ☊	☿	♀	♂	⚳	♃	♄	♅	♆	♇
1 Su	22 39 28	8♍17 10	0♎32 05	8♎05 10	16♋20.0	5♍17.1	13♍09.5	8♍45.9	5♐58.4	15♐08.4	14♑09.1	6♉27.3	17♓33.9	20♑52.7
2 M	22 43 25	9 15 14	15 33 54	22 57 15	16R 12.2	7 14.6	14 23.9	9 24.1	6 11.8	15 12.2	14R 07.5	6R 26.3	17R 32.3	20R 51.8
3 Tu	22 47 22	10 13 19	0♏14 27	7♏24 57	16 05.5	9 11.4	15 38.3	10 02.3	6 25.4	15 16.2	14 06.0	6 25.3	17 30.7	20 51.0
4 W	22 51 18	11 11 26	14 28 22	21 24 36	16 00.5	11 07.4	16 52.7	10 40.6	6 39.2	15 20.3	14 04.6	6 24.2	17 29.0	20 50.2
5 Th	22 55 15	12 09 34	28 13 40	4♐55 47	15 57.6	13 02.4	18 07.2	11 18.8	6 53.3	15 24.6	14 03.2	6 23.1	17 27.4	20 49.4
6 F	22 59 11	13 07 44	11♐31 18	18 00 36	15D 56.7	14 56.4	19 21.6	11 57.1	7 07.5	15 29.0	14 02.0	6 21.9	17 25.7	20 48.6
7 Sa	23 03 08	14 05 55	24 24 14	0♑42 42	15 57.1	16 49.3	20 36.0	12 35.3	7 22.0	15 33.7	14 00.8	6 20.7	17 24.1	20 47.9
8 Su	23 07 04	15 04 07	6♑56 36	13 06 30	15R 57.9	18 41.2	21 50.5	13 13.6	7 36.6	15 38.5	13 59.8	6 19.5	17 22.4	20 47.2
9 M	23 11 01	16 02 22	19 12 56	25 16 29	15 58.1	20 31.9	23 05.0	13 51.9	7 51.4	15 43.4	13 58.8	6 18.2	17 20.8	20 46.5
10 Tu	23 14 57	17 00 37	1♒17 38	7♒16 52	15 56.8	22 21.5	24 19.4	14 30.2	8 06.4	15 48.5	13 58.0	6 16.9	17 19.1	20 45.8
11 W	23 18 54	17 58 54	13 14 38	19 11 19	15 53.3	24 09.9	25 33.9	15 08.5	8 21.6	15 53.8	13 57.2	6 15.5	17 17.5	20 45.2
12 Th	23 22 51	18 57 13	25 07 17	1♓02 49	15 47.2	25 57.2	26 48.4	15 46.8	8 37.0	15 59.3	13 56.6	6 14.1	17 15.8	20 44.6
13 F	23 26 47	19 55 34	6♓58 13	12 53 44	15 38.5	27 43.4	28 02.9	16 25.1	8 52.6	16 04.9	13 56.0	6 12.7	17 14.2	20 44.0
14 Sa	23 30 44	20 53 56	18 49 33	24 45 53	15 27.7	29 28.4	29 17.4	17 03.5	9 08.3	16 10.6	13 55.5	6 11.2	17 12.5	20 43.4
15 Su	23 34 40	21 52 20	0♈42 53	6♈40 46	15 15.4	1♎12.4	0♎31.9	17 41.8	9 24.2	16 16.5	13 55.1	6 09.6	17 10.9	20 42.9
16 M	23 38 37	22 50 46	12 39 40	18 39 47	15 02.8	2 55.2	1 46.4	18 20.2	9 40.3	16 22.6	13 54.9	6 08.1	17 09.2	20 42.4
17 Tu	23 42 33	23 49 14	24 41 19	0♉44 29	14 50.9	4 37.0	3 00.9	18 58.5	9 56.5	16 28.8	13 54.7	6 06.5	17 07.6	20 41.9
18 W	23 46 30	24 47 44	6♉49 32	12 56 46	14 40.7	6 17.7	4 15.4	19 36.9	10 12.9	16 35.1	13D 54.6	6 04.8	17 06.0	20 41.4
19 Th	23 50 26	25 46 16	19 06 30	25 19 05	14 32.9	7 57.4	5 29.9	20 15.3	10 29.5	16 41.6	13 54.6	6 03.1	17 04.3	20 41.0
20 F	23 54 23	26 44 51	1♊34 54	7♊54 25	14 27.8	9 36.0	6 44.4	20 53.7	10 46.2	16 48.3	13 54.7	6 01.4	17 02.7	20 40.6
21 Sa	23 58 19	27 43 27	14 18 02	20 46 14	14 25.2	11 13.7	7 59.0	21 32.1	11 03.1	16 55.1	13 54.9	5 59.7	17 01.1	20 40.2
22 Su	0 02 16	28 42 06	27 19 27	3♋58 09	14D 24.5	12 50.3	9 13.5	22 10.6	11 20.1	17 02.1	13 55.2	5 57.9	16 59.5	20 39.9
23 M	0 06 13	29 40 47	10♋42 43	17 33 29	14R 24.6	14 26.0	10 28.1	22 49.0	11 37.3	17 09.2	13 55.6	5 56.1	16 57.9	20 39.6
24 Tu	0 10 09	0♎39 30	24 30 41	1♌34 26	14 24.4	16 00.7	11 42.6	23 27.5	11 54.7	17 16.4	13 56.2	5 54.2	16 56.3	20 39.3
25 W	0 14 06	1 38 16	8♌44 42	16 01 14	14 22.6	17 34.4	12 57.2	24 06.0	12 12.2	17 23.8	13 56.8	5 52.3	16 54.7	20 39.1
26 Th	0 18 02	2 37 04	23 23 38	0♍51 15	14 18.3	19 07.2	14 11.8	24 44.5	12 29.8	17 31.3	13 57.5	5 50.4	16 53.1	20 38.8
27 F	0 21 59	3 35 54	8♍23 14	15 58 29	14 11.4	20 39.1	15 26.3	25 23.0	12 47.6	17 38.9	13 58.3	5 48.5	16 51.5	20 38.6
28 Sa	0 25 55	4 34 46	23 35 47	1♎13 47	14 01.9	22 10.0	16 40.9	26 01.5	13 05.5	17 46.7	13 59.2	5 46.5	16 50.0	20 38.4
29 Su	0 29 52	5 33 40	8♎51 02	16 26 09	13 51.0	23 40.0	17 55.5	26 40.0	13 23.6	17 54.7	14 00.2	5 44.5	16 48.4	20 38.3
30 M	0 33 48	6 32 36	23 57 46	1♏24 42	13 39.8	25 09.0	19 10.1	27 18.6	13 41.8	18 02.7	14 01.3	5 42.4	16 46.9	20 38.2

LONGITUDE — October 2019

Day	Sid.Time	☉	0 hr ☽	Noon ☽	True ☊	☿	♀	♂	⚳	♃	♄	♅	♆	♇
1 Tu	0 37 45	7♎31 34	8♏45 55	16♏00 37	13♋29.7	26♎37.2	20♎24.7	27♍57.2	14♐00.2	18♐10.9	14♑02.5	5♉40.4	16♓45.3	20♑38.1
2 W	0 41 42	8 30 34	23 08 13	0♐08 24	13R 21.7	28 04.3	21 39.3	28 35.7	14 18.6	18 19.2	14 03.8	5R 38.3	16R 43.8	20R 38.0
3 Th	0 45 38	9 29 36	7♐01 02	13 46 11	13 16.3	29 30.6	22 53.9	29 14.3	14 37.3	18 27.7	14 05.1	5 36.1	16 42.3	20D 38.0
4 F	0 49 35	10 28 39	20 24 07	26 55 13	13 13.4	0♏55.8	24 08.5	29 52.9	14 56.0	18 36.3	14 06.6	5 34.0	16 40.8	20 38.0
5 Sa	0 53 31	11 27 45	3♑19 57	9♑38 54	13 12.5	2 20.1	25 23.0	0♎31.6	15 14.9	18 45.0	14 08.2	5 31.8	16 39.4	20 38.1
6 Su	0 57 28	12 26 52	15 52 41	22 01 58	13 12.4	3 43.3	26 37.6	1 10.2	15 33.8	18 53.8	14 09.9	5 29.6	16 37.9	20 38.1
7 M	1 01 24	13 26 01	28 07 22	4♒09 34	13 12.0	5 05.4	27 52.2	1 48.9	15 53.0	19 02.7	14 11.7	5 27.4	16 36.5	20 38.2
8 Tu	1 05 21	14 25 11	10♒09 12	16 06 52	13 10.1	6 26.5	29 06.8	2 27.5	16 12.2	19 11.8	14 13.6	5 25.2	16 35.1	20 38.4
9 W	1 09 17	15 24 24	22 03 08	27 58 31	13 06.0	7 46.4	0♏21.4	3 06.2	16 31.5	19 21.0	14 15.6	5 22.9	16 33.7	20 38.5
10 Th	1 13 14	16 23 38	3♓53 31	9♓48 33	12 59.0	9 05.2	1 36.0	3 44.9	16 51.0	19 30.3	14 17.6	5 20.7	16 32.3	20 38.7
11 F	1 17 11	17 22 54	15 44 00	21 40 10	12 49.1	10 22.6	2 50.5	4 23.6	17 10.5	19 39.7	14 19.8	5 18.4	16 30.9	20 38.9
12 Sa	1 21 07	18 22 12	27 37 21	3♈35 47	12 36.9	11 38.8	4 05.1	5 02.3	17 30.2	19 49.2	14 22.0	5 16.0	16 29.6	20 39.1
13 Su	1 25 04	19 21 32	9♈35 39	15 37 05	12 23.1	12 53.5	5 19.7	5 41.0	17 50.0	19 58.8	14 24.4	5 13.7	16 28.2	20 39.4
14 M	1 29 00	20 20 54	21 40 15	27 45 13	12 08.9	14 06.6	6 34.3	6 19.8	18 09.9	20 08.6	14 26.8	5 11.4	16 26.9	20 39.7
15 Tu	1 32 57	21 20 18	3♉52 06	10♉00 59	11 55.3	15 18.2	7 48.8	6 58.6	18 29.9	20 18.4	14 29.3	5 09.0	16 25.6	20 40.1
16 W	1 36 53	22 19 44	16 11 59	22 25 13	11 43.6	16 28.0	9 03.4	7 37.4	18 50.0	20 28.4	14 32.0	5 06.6	16 24.4	20 40.4
17 Th	1 40 50	23 19 12	28 40 49	4♊58 58	11 34.6	17 35.8	10 18.0	8 16.2	19 10.2	20 38.5	14 34.7	5 04.2	16 23.1	20 40.8
18 F	1 44 46	24 18 43	11♊19 51	17 43 42	11 28.5	18 41.6	11 32.5	8 55.0	19 30.4	20 48.6	14 37.5	5 01.8	16 21.9	20 41.2
19 Sa	1 48 43	25 18 15	24 10 49	0♋41 29	11 25.2	19 45.2	12 47.1	9 33.8	19 50.8	20 58.9	14 40.4	4 59.4	16 20.7	20 41.7
20 Su	1 52 39	26 17 50	7♋16 00	13 54 42	11D 24.1	20 46.2	14 01.7	10 12.7	20 11.3	21 09.3	14 43.4	4 57.0	16 19.5	20 42.2
21 M	1 56 36	27 17 28	20 37 55	27 25 54	11R 24.2	21 44.5	15 16.2	10 51.6	20 31.9	21 19.8	14 46.4	4 54.6	16 18.3	20 42.7
22 Tu	2 00 33	28 17 08	4♌18 54	11♌17 06	11 24.2	22 39.8	16 30.8	11 30.5	20 52.6	21 30.3	14 49.6	4 52.1	16 17.2	20 43.2
23 W	2 04 29	29 16 49	18 20 32	25 29 09	11 22.8	23 31.7	17 45.4	12 09.4	21 13.4	21 41.0	14 52.9	4 49.7	16 16.1	20 43.8
24 Th	2 08 26	0♏16 34	2♍42 44	10♍00 55	11 19.3	24 20.0	19 00.0	12 48.4	21 34.3	21 51.8	14 56.2	4 47.2	16 15.0	20 44.4
25 F	2 12 22	1 16 20	17 23 07	24 48 35	11 13.1	25 04.2	20 14.5	13 27.3	21 55.2	22 02.6	14 59.6	4 44.8	16 14.0	20 45.0
26 Sa	2 16 19	2 16 09	2♎16 27	9♎45 39	11 04.5	25 43.9	21 29.1	14 06.3	22 16.3	22 13.6	15 03.1	4 42.3	16 12.9	20 45.7
27 Su	2 20 15	3 15 59	17 15 01	24 43 23	10 54.4	26 18.6	22 43.7	14 45.3	22 37.4	22 24.6	15 06.7	4 39.8	16 11.9	20 46.4
28 M	2 24 12	4 15 52	2♏09 32	9♏32 20	10 43.9	26 47.8	23 58.3	15 24.3	22 58.6	22 35.8	15 10.4	4 37.4	16 10.9	20 47.1
29 Tu	2 28 08	5 15 47	16 50 44	24 03 52	10 34.2	27 10.9	25 12.8	16 03.4	23 19.9	22 47.0	15 14.2	4 34.9	16 10.0	20 47.8
30 W	2 32 05	6 15 44	1♐11 01	8♐11 40	10 26.5	27 27.4	26 27.4	16 42.4	23 41.3	22 58.3	15 18.0	4 32.4	16 09.0	20 48.6
31 Th	2 36 02	7 15 42	15 05 30	21 52 23	10 21.2	27R 36.5	27 42.0	17 21.5	24 02.7	23 09.7	15 21.9	4 30.0	16 08.2	20 49.4

Astro Data (Dy Hr Mn)	Planet Ingress (Dy Hr Mn)	Last Aspect (Dy Hr Mn)	☽ Ingress (Dy Hr Mn)	Last Aspect (Dy Hr Mn)	☽ Ingress (Dy Hr Mn)	☽ Phases & Eclipses (Dy Hr Mn)
☽0S 1 18:02	☿ ♎ 14 7:16	2 8:35 ♇ □	♏ 2 23:36	2 9:47 ♂ ⚹	♐ 2 11:45	6 3:12 ☽ 13♐15
☊D 6 2:58	♀ ♎ 14 13:44	4 10:59 ♇ ⚹	♐ 5 3:09	4 7:35 ♀ ⚹	♑ 4 17:44	14 4:34 ○ 21♓05
☊R 8 17:38	☉ ♎ 23 7:51	6 16:04 ♀ □	♑ 7 10:38	6 23:27 ♀ □	♒ 7 3:43	22 2:42 ☾ 28♊49
☿0S 15 10:28		9 8:31 ♀ △	♒ 9 21:25	8 18:28 ♃ ⚹	♓ 9 16:06	28 18:28 ● 5♎20
☽0N 15 22:03	☿ ♏ 3 8:15	11 5:24 ♃ ⚹	♓ 12 9:53	11 9:56 ♇ ⚹	♈ 12 4:47	
♀0S 16 22:08	♂ ♎ 4 4:23	14 4:34 ☉ ☍	♈ 14 22:34	13 22:00 ♇ □	♉ 14 16:25	5 16:48 ☽ 12♑09
♄D 18 8:48	♀ ♏ 8 17:07	16 16:04 ♇ □	♉ 17 10:32	16 8:39 ♇ △	♊ 17 2:31	13 21:09 ○ 20♈14
♃□♆ 21 16:45	☉ ♏ 23 17:21	19 13:58 ☉ △	♊ 19 20:59	19 2:15 ☉ △	♋ 19 10:44	21 12:40 ☾ 27♋49
☊D 22 2:46		22 2:42 ☉ □	♋ 22 4:51	21 12:40 ☉ □	♌ 21 16:30	28 3:40 ● 4♏25
☊R 23 6:33	♃♀♀14 5:27	23 22:06 ♂ ⚹	♌ 24 9:21	23 9:15 ☿ □	♍ 23 19:31	
☉0S 23 7:51	♃⚺⚺17 5:46	25 16:15 ☿ ⚹	♍ 26 10:38	25 13:01 ☿ ⚹	♎ 25 20:21	
☽0S 29 4:26	☊ D20 7:27	28 3:59 ♂ ☌	♎ 28 10:04	27 8:23 ♃ ⚹	♏ 27 20:30	
♇D 3 6:41	☊ R21 12:52	30 2:07 ☿ ☌	♏ 30 9:43	29 17:36 ☿ ☌	♐ 29 22:00	
♂0S 7 14:19	☽0S26 15:09					
☽0N 13 4:03	☿ R31 15:43					

Astro Data

1 September 2019
Julian Day # 43708
SVP 4♓59'22"
GC 27♐06.8 ⚴ 2♏04.2
Eris 24♈10.9R ⚵ 3♍35.0
⚷ 4♈46.8R ⚶ 25♉49.8
☽ Mean ☊ 14♋41.9

1 October 2019
Julian Day # 43738
SVP 4♓59'20"
GC 27♐06.9 ⚴ 13♏49.9
Eris 23♈55.9R ⚵ 16♍31.9
⚷ 3♈28.6R ⚶ 27♉28.7R
☽ Mean ☊ 13♋06.6

*Giving the positions of planets daily at midnight, Greenwich Mean Time (0:00 UT)
Each planet's retrograde period is shaded gray.

2019 Planetary Ephemeris

November 2019 — LONGITUDE

Day	Sid.Time	☉	0 hr ☽	Noon ☽	True ☊	☿	♀	♂	⚳	♃	♄	♅	♆	♇
1 F	2 39 58	8♏15 43	28♐32 23	5♑05 41	10♋18.5	27♏37.8	28♏56.5	18♎00.6	24♐24.3	23♐21.2	15♑25.9	4♉27.5	16♓07.3	20♑50.2
2 Sa	2 43 55	9 15 45	11♑32 38	17 53 41	10D 17.9	27R 30.5	0♐11.1	18 39.7	24 45.9	23 32.7	15 30.0	4R 25.1	16R 06.4	20 51.1
3 Su	2 47 51	10 15 48	24 09 20	0♒20 12	10 18.4	27 14.3	1 25.6	19 18.8	25 07.6	23 44.3	15 34.2	4 22.6	16 05.6	20 52.0
4 M	2 51 48	11 15 53	6♒26 54	12 30 05	10R 19.2	26 48.6	2 40.2	19 58.0	25 29.4	23 56.1	15 38.4	4 20.2	16 04.9	20 52.9
5 Tu	2 55 44	12 16 00	18 30 24	24 28 32	10 19.0	26 13.2	3 54.7	20 37.1	25 51.2	24 07.8	15 42.8	4 17.7	16 04.1	20 53.8
6 W	2 59 41	13 16 08	0♓25 07	6♓20 46	10 17.2	25 28.3	5 09.3	21 16.3	26 13.1	24 19.7	15 47.1	4 15.3	16 03.4	20 54.8
7 Th	3 03 37	14 16 18	12 16 04	18 11 33	10 13.2	24 34.1	6 23.8	21 55.5	26 35.1	24 31.6	15 51.6	4 12.9	16 02.7	20 55.8
8 F	3 07 34	15 16 30	24 07 44	0♈05 05	10 06.8	23 31.4	7 38.3	22 34.7	26 57.1	24 43.7	15 56.2	4 10.5	16 02.0	20 56.8
9 Sa	3 11 31	16 16 42	6♈03 58	12 04 44	9 58.5	22 21.5	8 52.8	23 13.9	27 19.2	24 55.7	16 00.8	4 08.1	16 01.4	20 57.8
10 Su	3 15 27	17 16 57	18 07 42	24 13 04	9 48.7	21 06.0	10 07.3	23 53.2	27 41.4	25 07.9	16 05.5	4 05.7	16 00.8	20 58.9
11 M	3 19 24	18 17 13	0♉21 01	6♉31 42	9 38.5	19 47.1	11 21.8	24 32.4	28 03.6	25 20.1	16 10.2	4 03.3	16 00.2	21 00.0
12 Tu	3 23 20	19 17 31	12 45 11	19 01 31	9 28.8	18 27.2	12 36.3	25 11.7	28 25.9	25 32.4	16 15.1	4 01.0	15 59.7	21 01.1
13 W	3 27 17	20 17 50	25 20 43	1♊42 47	9 20.5	17 08.8	13 50.8	25 51.0	28 48.2	25 44.7	16 20.0	3 58.6	15 59.2	21 02.3
14 Th	3 31 13	21 18 12	8♊07 42	14 35 26	9 14.2	15 54.5	15 05.2	26 30.4	29 10.7	25 57.1	16 24.9	3 56.3	15 58.7	21 03.5
15 F	3 35 10	22 18 35	21 05 59	27 39 20	9 10.3	14 46.7	16 19.7	27 09.7	29 33.1	26 09.6	16 30.0	3 54.0	15 58.3	21 04.7
16 Sa	3 39 06	23 18 59	4♋15 31	10♋54 32	9D 08.7	13 47.5	17 34.2	27 49.1	29 55.7	26 22.1	16 35.1	3 51.7	15 57.9	21 05.9
17 Su	3 43 03	24 19 26	17 36 27	24 21 19	9 08.8	12 58.2	18 48.6	28 28.5	0♑18.2	26 34.7	16 40.2	3 49.4	15 57.5	21 07.1
18 M	3 47 00	25 19 55	1♌09 13	8♌00 14	9 10.0	12 20.1	20 03.1	29 07.9	0 40.9	26 47.4	16 45.5	3 47.2	15 57.1	21 08.4
19 Tu	3 50 56	26 20 25	14 54 23	21 51 44	9R 11.3	11 53.6	21 17.5	29 47.4	1 03.6	27 00.1	16 50.8	3 45.0	15 56.8	21 09.7
20 W	3 54 53	27 20 57	28 52 14	5♍55 50	9 11.8	11D 38.8	22 32.0	0♏26.8	1 26.3	27 12.8	16 56.1	3 42.7	15 56.6	21 11.0
21 Th	3 58 49	28 21 31	13♍02 23	20 11 37	9 10.8	11 35.4	23 46.4	1 06.3	1 49.1	27 25.6	17 01.6	3 40.6	15 56.3	21 12.4
22 F	4 02 46	29 22 07	27 23 13	4♎36 43	9 08.0	11 42.9	25 00.8	1 45.8	2 12.0	27 38.5	17 07.0	3 38.4	15 56.1	21 13.8
23 Sa	4 06 42	0♐22 44	11♎51 35	19 07 11	9 03.6	12 00.6	26 15.2	2 25.3	2 34.9	27 51.4	17 12.6	3 36.3	15 55.9	21 15.1
24 Su	4 10 39	1 23 23	26 22 48	3♏37 39	8 57.9	12 27.6	27 29.7	3 04.9	2 57.8	28 04.4	17 18.2	3 34.2	15 55.8	21 16.6
25 M	4 14 35	2 24 04	10♏50 57	18 01 54	8 51.9	13 03.1	28 44.1	3 44.5	3 20.8	28 17.4	17 23.8	3 32.1	15 55.7	21 18.0
26 Tu	4 18 32	3 24 46	25 09 43	2♐13 44	8 46.3	13 46.0	29 58.5	4 24.1	3 43.8	28 30.5	17 29.6	3 30.0	15 55.6	21 19.5
27 W	4 22 29	4 25 30	9♐13 20	16 08 01	8 41.9	14 35.7	1♑12.9	5 03.7	4 06.9	28 43.6	17 35.3	3 28.0	15D 55.6	21 20.9
28 Th	4 26 25	5 26 15	22 57 24	29 41 14	8 39.2	15 31.2	2 27.3	5 43.3	4 30.1	28 56.8	17 41.2	3 26.0	15 55.6	21 22.4
29 F	4 30 22	6 27 02	6♑19 25	12♑51 58	8D 38.1	16 31.9	3 41.6	6 23.0	4 53.2	29 10.0	17 47.0	3 24.1	15 55.6	21 24.0
30 Sa	4 34 18	7 27 49	19 18 59	25 40 44	8 38.5	17 37.0	4 56.0	7 02.6	5 16.5	29 23.2	17 53.0	3 22.1	15 55.7	21 25.5

December 2019 — LONGITUDE

Day	Sid.Time	☉	0 hr ☽	Noon ☽	True ☊	☿	♀	♂	⚳	♃	♄	♅	♆	♇
1 Su	4 38 15	8♐28 38	1♒57 32	8♒09 47	8♋39.9	18♏45.9	6♑10.3	7♏42.3	5♑39.7	29♐36.5	17♑59.0	3♉20.2	15♓55.8	21♑27.1
2 M	4 42 11	9 29 27	14 17 59	20 22 38	8 41.8	19 58.2	7 24.7	8 22.0	6 03.0	29 49.8	18 05.0	3R 18.4	15 55.9	21 28.6
3 Tu	4 46 08	10 30 18	26 24 19	2♓23 38	8 43.4	21 13.4	8 39.0	9 01.8	6 26.3	0♑03.1	18 11.1	3 16.6	15 56.1	21 30.2
4 W	4 50 05	11 31 09	8♓21 11	14 17 35	8R 44.2	22 31.0	9 53.3	9 41.5	6 49.7	0 16.5	18 17.2	3 14.8	15 56.3	21 31.9
5 Th	4 54 01	12 32 01	20 13 29	26 09 28	8 43.9	23 50.7	11 07.6	10 21.3	7 13.1	0 30.0	18 23.4	3 13.0	15 56.5	21 33.5
6 F	4 57 58	13 32 54	2♈06 08	8♈04 03	8 42.5	25 12.2	12 21.8	11 01.0	7 36.5	0 43.4	18 29.6	3 11.3	15 56.8	21 35.2
7 Sa	5 01 54	14 33 48	14 03 43	20 05 40	8 39.9	26 35.2	13 36.1	11 40.9	8 00.0	0 56.9	18 35.9	3 09.6	15 57.1	21 36.8
8 Su	5 05 51	15 34 42	26 10 19	2♉18 04	8 36.5	27 59.6	14 50.3	12 20.7	8 23.5	1 10.4	18 42.2	3 08.0	15 57.5	21 38.5
9 M	5 09 47	16 35 37	8♉29 15	14 44 08	8 32.8	29 25.1	16 04.5	13 00.5	8 47.0	1 23.9	18 48.5	3 06.4	15 57.8	21 40.2
10 Tu	5 13 44	17 36 34	21 02 54	27 25 42	8 29.2	0♐51.6	17 18.7	13 40.4	9 10.6	1 37.5	18 54.9	3 04.8	15 58.3	21 41.9
11 W	5 17 40	18 37 31	3♊52 36	10♊23 34	8 26.2	2 18.9	18 32.9	14 20.3	9 34.1	1 51.1	19 01.4	3 03.3	15 58.7	21 43.7
12 Th	5 21 37	19 38 29	16 58 34	23 37 28	8 24.1	3 46.9	19 47.0	15 00.2	9 57.7	2 04.7	19 07.8	3 01.8	15 59.2	21 45.4
13 F	5 25 34	20 39 27	0♋20 05	7♋06 11	8D 23.0	5 15.5	21 01.1	15 40.1	10 21.4	2 18.4	19 14.3	3 00.4	15 59.7	21 47.2
14 Sa	5 29 30	21 40 27	13 55 31	20 47 48	8 22.9	6 44.7	22 15.2	16 20.1	10 45.0	2 32.0	19 20.9	2 59.0	16 00.3	21 49.0
15 Su	5 33 27	22 41 27	27 42 44	4♌40 01	8 23.6	8 14.3	23 29.3	17 00.0	11 08.7	2 45.7	19 27.5	2 57.6	16 00.8	21 50.8
16 M	5 37 23	23 42 29	11♌39 21	18 40 25	8 24.7	9 44.4	24 43.4	17 40.1	11 32.4	2 59.4	19 34.1	2 56.3	16 01.5	21 52.6
17 Tu	5 41 20	24 43 31	25 42 57	2♍46 38	8 25.9	11 14.8	25 57.4	18 20.1	11 56.2	3 13.1	19 40.7	2 55.0	16 02.1	21 54.4
18 W	5 45 16	25 44 34	9♍51 12	16 56 24	8 26.8	12 45.5	27 11.4	19 00.1	12 19.9	3 26.9	19 47.4	2 53.8	16 02.8	21 56.3
19 Th	5 49 13	26 45 39	24 01 58	1♎07 38	8R 27.2	14 16.6	28 25.4	19 40.2	12 43.7	3 40.6	19 54.1	2 52.6	16 03.5	21 58.1
20 F	5 53 09	27 46 44	8♎13 08	15 18 10	8 27.1	15 47.9	29 39.3	20 20.3	13 07.5	3 54.4	20 00.9	2 51.5	16 04.3	22 00.0
21 Sa	5 57 06	28 47 50	22 22 28	29 25 43	8 26.4	17 19.4	0♒53.3	21 00.4	13 31.3	4 08.2	20 07.6	2 50.4	16 05.1	22 01.9
22 Su	6 01 03	29 48 56	6♏27 36	13♏27 47	8 25.5	18 51.3	2 07.2	21 40.5	13 55.1	4 22.0	20 14.4	2 49.4	16 05.9	22 03.7
23 M	6 04 59	0♑50 04	20 25 56	27 21 42	8 24.4	20 23.3	3 21.0	22 20.7	14 19.0	4 35.8	20 21.2	2 48.4	16 06.7	22 05.6
24 Tu	6 08 56	1 51 12	4♐14 45	11♐04 46	8 23.6	21 55.6	4 34.9	23 00.9	14 42.9	4 49.6	20 28.1	2 47.4	16 07.6	22 07.6
25 W	6 12 52	2 52 21	17 51 27	24 34 33	8 23.0	23 28.2	5 48.7	23 41.1	15 06.8	5 03.4	20 35.0	2 46.5	16 08.5	22 09.5
26 Th	6 16 49	3 53 31	1♑13 49	7♑49 06	8D 22.7	25 01.0	7 02.5	24 21.3	15 30.7	5 17.2	20 41.9	2 45.7	16 09.5	22 11.4
27 F	6 20 45	4 54 41	14 20 18	20 47 20	8 22.7	26 34.0	8 16.3	25 01.5	15 54.6	5 31.1	20 48.8	2 44.9	16 10.5	22 13.3
28 Sa	6 24 42	5 55 51	27 10 14	3♒29 04	8 22.8	28 07.2	9 30.0	25 41.8	16 18.5	5 44.9	20 55.7	2 44.1	16 11.5	22 15.3
29 Su	6 28 38	6 57 01	9♒44 01	15 55 15	8 23.0	29 40.7	10 43.7	26 22.1	16 42.5	5 58.7	21 02.7	2 43.4	16 12.5	22 17.2
30 M	6 32 35	7 58 11	22 03 05	28 07 50	8R 23.1	1♑14.5	11 57.3	27 02.4	17 06.4	6 12.6	21 09.7	2 42.8	16 13.6	22 19.2
31 Tu	6 36 32	8 59 21	4♓09 53	10♓09 42	8 23.1	2 48.6	13 10.9	27 42.7	17 30.4	6 26.4	21 16.7	2 42.2	16 14.7	22 21.2

Astro Data	Dy Hr Mn
☊ D	1 21:39
☊ R	4 10:16
♄⚹♆	9 2:47
☽ 0N	9 11:33
☊ D	16 8:48
☊ R	19 21:23
☿ D	20 19:13
☽ 0S	23 0:16
♆ D	27 12:33
☊ D	29 4:11
☊ R	4 6:51
☽ 0N	6 20:14
☊ D	13 14:11
♃△♅	15 19:02
☊ R	19 5:08

Planet Ingress	Dy Hr Mn
♀ ♐	1 20:26
⚳ ♑	16 4:37
♂ ♏	19 7:41
☉ ♐	22 15:00
♀ ♑	26 0:30
♃ ♑	2 18:21
☿ ♐	9 9:43
♀ ♒	20 6:43
☉ ♑	22 4:21
☿ ♑	29 4:56
☽ 0S20	6:52
☊ D26	12:46
☊ R30	10:18

Last Aspect Dy Hr Mn		☽ Ingress	Dy Hr Mn
31 14:31	♃ ☌	♑	1 2:39
3 5:48	☿ ⚹	♒	3 11:21
5 14:38	☿ □	♓	5 23:09
8 1:14	♃ □	♈	8 11:50
10 14:02	♃ △	♉	10 23:19
12 15:49	♇ △	♊	13 8:47
15 11:41	♂ △	♋	15 16:16
17 20:16	♂ □	♌	17 21:58
19 21:12	☉ □	♍	20 1:56
22 3:33	☉ ⚹	♎	22 4:21
24 2:51	♃ ⚹	♏	24 5:59
25 17:31	♇ ⚹	♐	26 8:12
28 10:51	♃ ☌	♑	28 12:34
30 3:58	♇ ☌	♒	30 20:14

Last Aspect Dy Hr Mn		☽ Ingress	Dy Hr Mn
2 12:28	☿ □	♓	3 7:12
5 8:16	☿ △	♈	5 19:46
7 15:03	♇ □	♉	8 7:30
10 1:14	♇ △	♊	10 16:48
12 5:13	☉ ☍	♋	12 23:24
14 15:58	♀ ☍	♌	15 3:57
16 22:11	☉ △	♍	17 7:17
19 8:08	♀ △	♎	19 10:06
21 11:47	☉ ⚹	♏	21 12:58
23 3:28	♂ ☌	♐	23 16:35
25 11:19	☿ ☌	♑	25 21:46
27 21:04	♂ ⚹	♒	28 5:22
30 10:25	♂ □	♓	30 15:43

☽ Phases & Eclipses Dy Hr Mn	
4 10:24	☽ 11♒42
12 13:36	○ 19♉52
19 21:12	☾ 27♌14
26 15:07	● 4♐03
4 6:59	☽ 11♓49
12 5:13	○ 19♊52
19 4:58	☾ 26♍58
26 5:14	● 4♑07
26 5:18:54	● A 03'39"

Astro Data

1 November 2019
Julian Day # 43769
SVP 4♓59'16"
GC 27♐07.0 ⚴ 26♏49.3
Eris 23♈37.6R ⚵ 28♍50.7
⚷ 2♈12.2R ⚶ 22♉34.9R
☽ Mean ☊ 11♋28.0

1 December 2019
Julian Day # 43799
SVP 4♓59'12"
GC 27♐07.0 ⚴ 9♐42.1
Eris 23♈22.3R ⚵ 9♎12.2
⚷ 1♈30.1R ⚶ 15♉09.9R
☽ Mean ☊ 9♋52.7

*Giving the positions of planets daily at midnight, Greenwich Mean Time (0:00 UT)
Each planet's retrograde period is shaded gray.

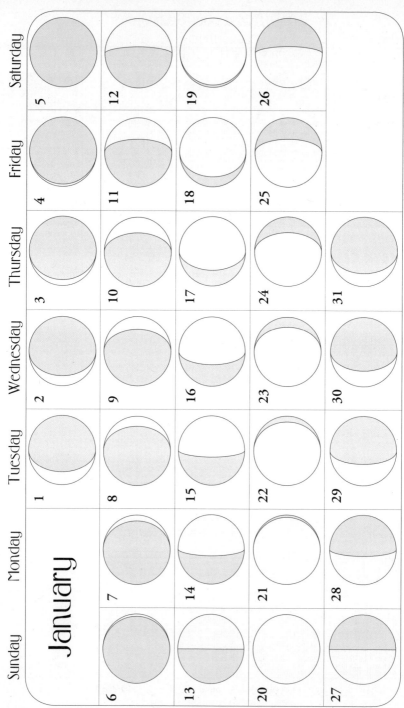

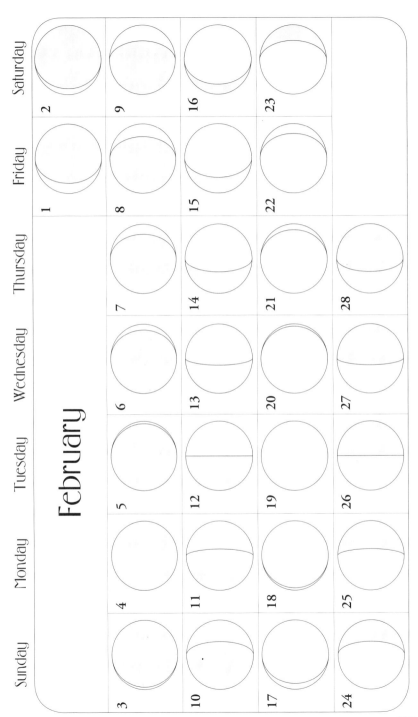

February

Sunday	Monday	Tuesday	Wednesday	Thursday	Friday	Saturday
					1	2
3	4	5	6	7	8	9
10	11	12	13	14	15	16
17	18	19	20	21	22	23
24	25	26	27	28		

215

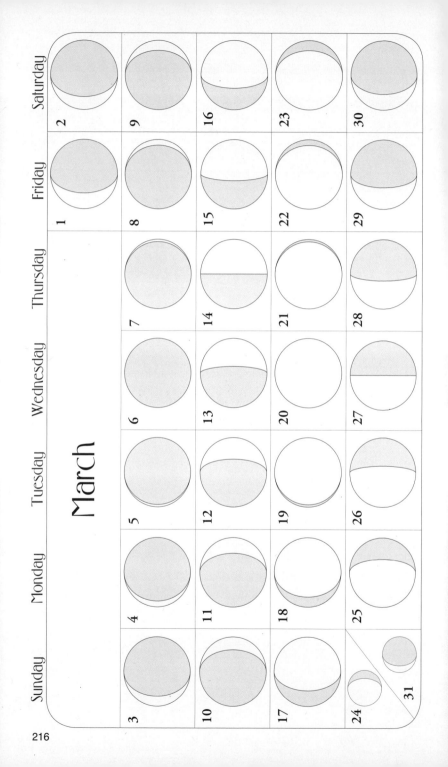

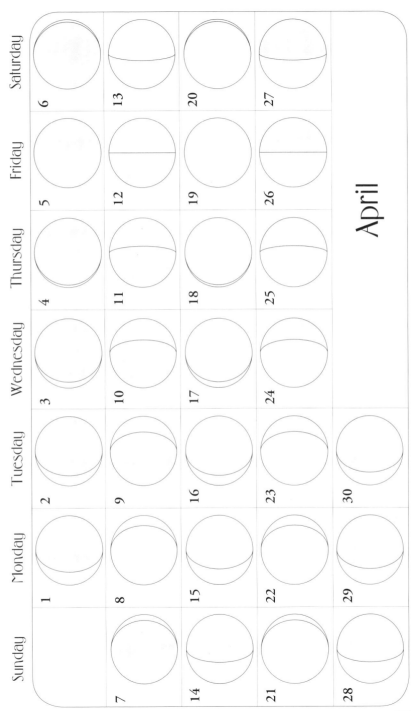

April

Sunday	Monday	Tuesday	Wednesday	Thursday	Friday	Saturday
	1	2	3	4	5	6
7	8	9	10	11	12	13
14	15	16	17	18	19	20
21	22	23	24	25	26	27
28	29	30				

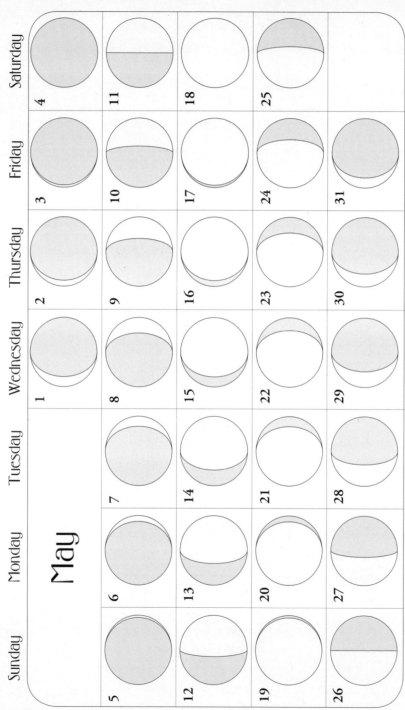

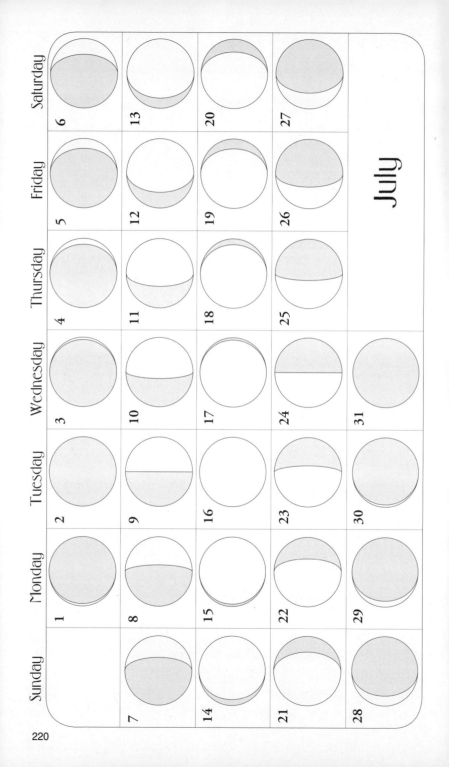

July

Sunday	Monday	Tuesday	Wednesday	Thursday	Friday	Saturday
	1	2	3	4	5	6
7	8	9	10	11	12	13
14	15	16	17	18	19	20
21	22	23	24	25	26	27
28	29	30	31			

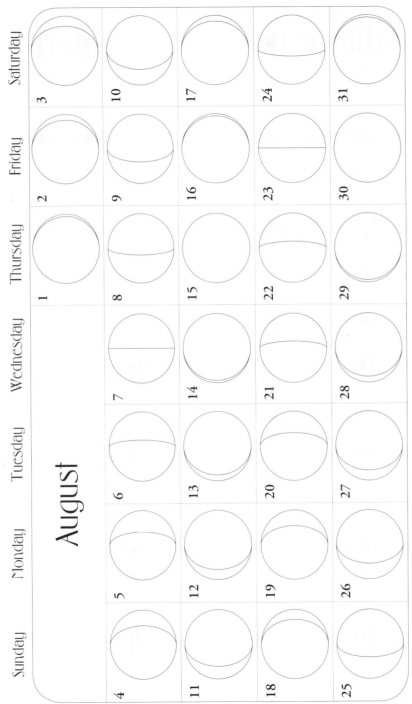

August

Sunday	Monday	Tuesday	Wednesday	Thursday	Friday	Saturday
				1	2	3
4	5	6	7	8	9	10
11	12	13	14	15	16	17
18	19	20	21	22	23	24
25	26	27	28	29	30	31

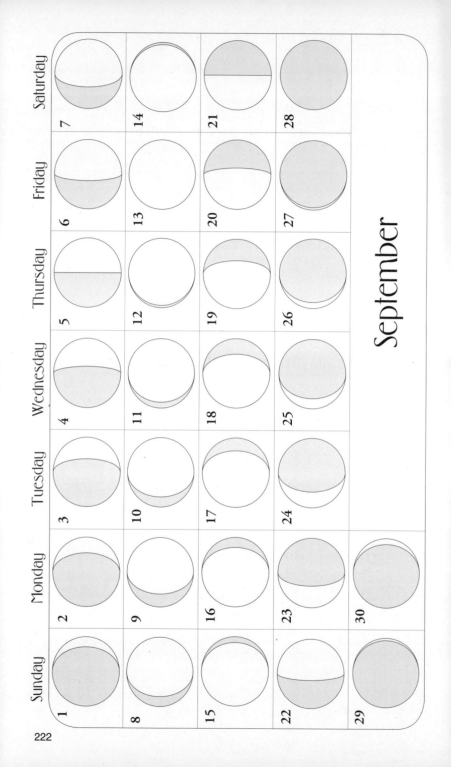

222

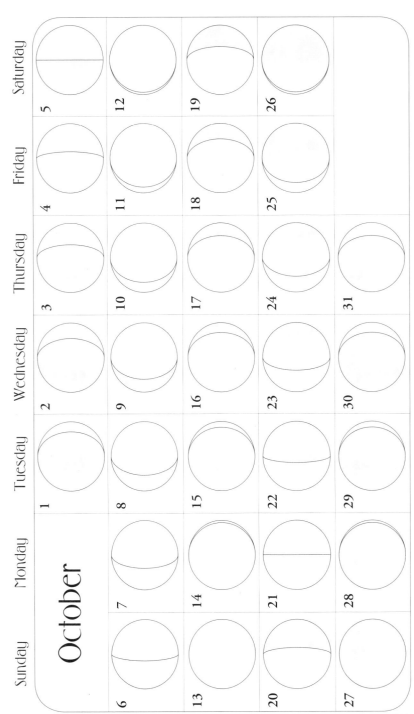

October

Sunday	Monday	Tuesday	Wednesday	Thursday	Friday	Saturday
		1	2	3	4	5
6	7	8	9	10	11	12
13	14	15	16	17	18	19
20	21	22	23	24	25	26
27	28	29	30	31		

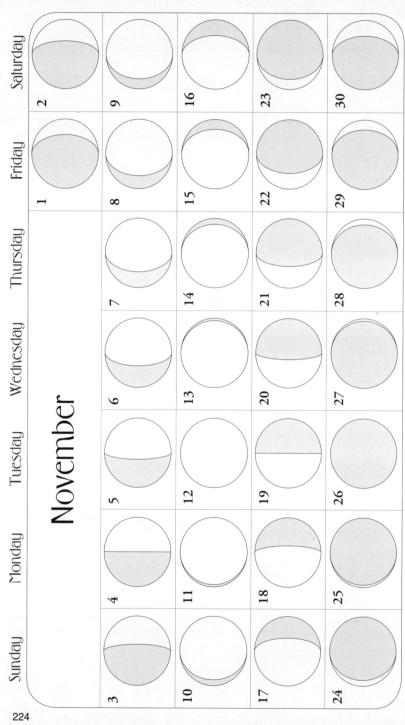

November

Sunday	Monday	Tuesday	Wednesday	Thursday	Friday	Saturday
					1	2
3	4	5	6	7	8	9
10	11	12	13	14	15	16
17	18	19	20	21	22	23
24	25	26	27	28	29	30

December

Eclipse Key:

☽ = Solar ☾ = Lunar **T**=Total **A**=Annular *n*=Penumbral **P**=Partial

Lunar Eclipses are visible wherever it is night and cloud free during full moon time.

Times on this page are in EST (Eastern Standard Time -5 from GMT)
or DST, Daylight Saving Time (Mar 10 - Nov 3, 2019)

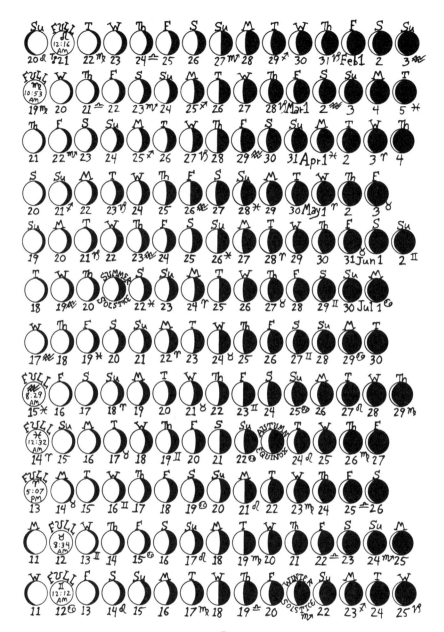

2019 Lunar Phases

This format available on cards from: **http://snakeandsnake.com**

Snake and Snake Productions 3037 Dixon Rd Durham, NC 27707

CONSTELLATIONS OF THE ZODIAC

These stations of the zodiac were named thousands of years ago for the constellations that were behind them at the time. The signs of the zodiac act like a light filter, coloring the qualities of life force. As the Sun, Moon and other planets move through the zodiac, the following influences are energized:

♒ **Aquarius** (Air): Community, ingenuity, collaboration, idealism. It's time to honor the philosophy of love and the power of community.

♓ **Pisces** (Water): Introspection, imagination, sensitivity and intuition. We process and gestate our dreams

♈ **Aries** (Fire): Brave, direct, rebellious, energized. Our inner teenager comes alive; our adult self needs to direct the energy wisely.

♉ **Taurus** (Earth): Sensual, rooted, nurturing, material manifestation. We slow down, get earthy, awaken our senses, begin to build form, roots, and stubborn strength.

♊ **Gemini** (Air): Communication, networking, curiosity, quick witted. We connect with like minds and build a network of understanding.

♋ **Cancer** (Water): Family, home, emotional awareness, nourishment. We need time in our shell and with our familiars.

♌ **Leo** (Fire): Creativity, charisma, warmth, and enthusiasm. Gather with others to celebrate and share bounty.

♍ **Virgo** (Earth): Mercurial, curious, critical, and engaged. The mood sharpens our minds and nerves, and sends us back to work.

♎ **Libra** (Air): Beauty, equality, egalitarianism, cooperation. We grow more friendly, relationship oriented, and incensed by injustice.

♏ **Scorpio** (Water): Sharp focus, perceptive, empowered, mysterious. The mood is smoky, primal, occult, and curious; still waters run deep.

♐ **Sagittarius** (Fire): Curiosity, honesty, exploration, playfulness. We grow more curious about what's unfamiliar.

♑ **Capricorn** (Earth): family, history, dreams, traditions. We need mountains to climb and problems to solve.

adapted from Heather Roan Robbins' Sun Signs and Sun Transits
© Mother Tongue Ink 2016

Signs and Symbols at a Glance

Planets

Personal Planets are closest to Earth.

⊙ **Sun**: self radiating outward, character, ego
☽ **Moon**: inward sense of self, emotions, psyche
☿ **Mercury**: communication, travel, thought
♀ **Venus**: relationship, love, sense of beauty, empathy
♂ **Mars**: will to act, initiative, ambition

Asteroids are between Mars and Jupiter and reflect the awakening of feminine-defined energy centers in human consciousness.

Social Planets are between personal and outer planets.

♃ **Jupiter**: expansion, opportunities, leadership
♄ **Saturn**: limits, structure, discipline
Note: The days of the week are named in various languages after the above seven heavenly bodies.
⚷ **Chiron**: is a small planetary body between Saturn and Uranus representing the wounded healer.

Transpersonal Planets are the outer planets.

♅ **Uranus**: cosmic consciousness, revolutionary change
♆ **Neptune**: spiritual awakening, cosmic love, all one
♇ **Pluto**: death and rebirth, deep, total change

Zodiac Signs

♈ Aries
♉ Taurus
♊ Gemini
♋ Cancer
♌ Leo
♍ Virgo
♎ Libra
♏ Scorpio
♐ Sagittarius
♑ Capricorn
♒ Aquarius
♓ Pisces

Aspects

Aspects show the angle between planets; this informs how the planets influence each other and us. **We'Moon** lists only significant aspects:

☌ CONJUNCTION (planets are 0–5° apart)
linked together, energy of aspected planets is mutually enhancing
☍ OPPOSITION (planets are 180° apart)
polarizing or complementing, energies are diametrically opposite
△ TRINE (planets are 120° apart)
harmonizing, energies of this aspect are in the same element
□ SQUARE (planets are 90° apart)
challenging, energies of this aspect are different from each other
⚹ SEXTILE (planets are 60° apart)
cooperative, energies of this aspect blend well
⚻ QUINCUNX (planets are 150° apart)
variable, energies of this aspect combine contrary elements

Other Symbols

☽ v/c–Moon is "void of course" from last lunar aspect until it enters new sign.
ApG–Apogee: Point in the orbit of the Moon that's farthest from Earth.
PrG–Perigee: Point in the orbit of the Moon that's nearest to Earth.
ApH–Aphelion: Point in the orbit of a planet that's farthest from the Sun.
PrH–Perihelion: Point in the orbit of a planet that's nearest to the Sun.
D or R–Direct or Retrograde: Describes when a planet moves forward (D) through the zodiac or appears to move backward (R).

2020

JANUARY
S	M	T	W	T	F	S
			1	2	3	4
5	6	7	8	9	10	11
12	13	14	15	16	17	18
19	20	21	22	23	24	25
26	27	28	29	30	31	

FEBRUARY
S	M	T	W	T	F	S
						1
2	3	4	5	6	7	8
9	10	11	12	13	14	15
16	17	18	19	20	21	22
23	24	25	26	27	28	29

MARCH
S	M	T	W	T	F	S
1	2	3	4	5	6	7
8	9	10	11	12	13	14
15	16	17	18	19	20	21
22	23	24	25	26	27	28
29	30	31				

APRIL
S	M	T	W	T	F	S
			1	2	3	4
5	6	7	8	9	10	11
12	13	14	15	16	17	18
19	20	21	22	23	24	25
26	27	28	29	30		

MAY
S	M	T	W	T	F	S
					1	2
3	4	5	6	7	8	9
10	11	12	13	14	15	16
17	18	19	20	21	22	23
24	25	26	27	28	29	30
31						

JUNE
S	M	T	W	T	F	S
	1	2	3	4	5	6
7	8	9	10	11	12	13
14	15	16	17	18	19	20
21	22	23	24	25	26	27
28	29	30				

JULY
S	M	T	W	T	F	S
			1	2	3	4
5	6	7	8	9	10	11
12	13	14	15	16	17	18
19	20	21	22	23	24	25
26	27	28	29	30	31	

AUGUST
S	M	T	W	T	F	S
						1
2	3	4	5	6	7	8
9	10	11	12	13	14	15
16	17	18	19	20	21	22
23	24	25	26	27	28	29
30	31					

SEPTEMBER
S	M	T	W	T	F	S
		1	2	3	4	5
6	7	8	9	10	11	12
13	14	15	16	17	18	19
20	21	22	23	24	25	26
27	28	29	30			

OCTOBER
S	M	T	W	T	F	S
				1	2	3
4	5	6	7	8	9	10
11	12	13	14	15	16	17
18	19	20	21	22	23	24
25	26	27	28	29	30	31

NOVEMBER
S	M	T	W	T	F	S
1	2	3	4	5	6	7
8	9	10	11	12	13	14
15	16	17	18	19	20	21
22	23	24	25	26	27	28
29	30					

DECEMBER
S	M	T	W	T	F	S
		1	2	3	4	5
6	7	8	9	10	11	12
13	14	15	16	17	18	19
20	21	22	23	24	25	26
27	28	29	30	31		

4 Seasons
© Serena Supplee 2010

● = NEW MOON, PST/PDT   ○ = FULL MOON, PST/PDT

230

PreacherWoman for the Goddess:
Poems, Invocations, Plays and Other Holy Writ

A spirit-filled word feast puzzling on life and death mysteries—rich with metaphor, surprise, earth-passion: a harvest of decades among women who build their own houses, bury their own dead, invent their own ceremonies.

By Bethroot Gwynn—poet, theaterwoman, temple keeper on women's land, and a longtime editor for the We'Moon Datebook.

Softbound, 6x9, 120 pages with 7 full color art features, $16

In the Spirit of We'Moon
Celebrating 30 Years:
An Anthology of We'Moon Art and Writing

This unique Anthology showcases three decades of We'Moon art, writing and herstory from 1981–2011. The anthology includes new insights from founding editor Musawa and other writers who share stories about We'Moon's colorful 30-year evolution. Now in its third printing!

256 full color pages, paperback, 8x10, $26.95

The Last Wild Witch
by Starhawk, illustrated by Lindy Kehoe

In the very heart of the last magic forest lived the last wild Witch...

In this story, the children of a perfect town found their joy and courage, and saved the last wild Witch and the forest from destruction. A Silver Nautilus Award winner, this book is recognized internationally for helping readers imagine a world as it could be with abundant possibilities.

34 full color pages, 8x10, Softbound, $9.95

An Eco-Fable for Kids and Other Free Spirits.

We'Moon 2019: Fanning the Flame

- **Datebook** The best-selling astrological moon calendar, earth-spirited handbook in natural rhythms, and visionary collection of women's creative work. Week-at-a-glance format. Choice of 3 bindings: Spiral, Sturdy Paperback Binding or Unbound. 8x5¼, 240 pages, $21.95

- *We'Moon en Español!*
We are proud to offer a full translation of the classic datebook, in Spanish! Spiral Bound, 240 pages, 8x5¼, $21.95

- **Cover Poster** featuring Emily Balivet's "*Brigid—The Goddess of Inspiration,*" sparking creativity and fiery will to act. 11x17, $10

- **We'Moon on the Wall**
A beautiful full color wall calendar featuring inspired art and writing from *We'Moon 2019,* with key astrological information, interpretive articles, lunar phases and signs. 12x12, $16.95

- **We'Moon 2019 Tote Bag** 100% Organic Cotton tote, proudly displaying the cover of *We'Moon 2019*. Perfect for stowing all of your goodies in style. 13x14, $13

• **Greeting Cards** An assortment of six gorgeous note cards featuring art from *We'Moon 2019*, with writings from each artist on the back. Wonderful to send for any occasion: Holy Day, Birthday, Anniversary, Sympathy, or just to say hello. Each pack is wrapped in biodegradable cellophane. Blank inside. 5x7, $11.95

More Offerings!
Check out page 231 for details on these books:

• *The Last Wild Witch* by Starhawk, illustrated by Lindy Kehoe.

• *In the Spirit of We'Moon ~ Celebrating 30 Years: An Anthology of We'Moon Art and Writing*

• *Preacher Woman for the Goddess: Poems, Invocations, Plays and Other Holy Writ* by We'Moon Special Editor Bethroot Gwynn.

ORDER NOW—WHILE THEY LAST!
Take advantage of our
Special Discounts:
• We'll ship orders of $50 or more for **FREE** within the US!
• Use promo code: **19Sun** to get 10% off
orders of $100 or more!
We often have great package deals and discounts.
Look for details, and sign up to receive regular email updates at
www.wemoon.ws
Email weorder@wemoon.ws Toll free in US 877-693-6666
Local & International 541-956-6052 Wholesale 503-288-3588
SHIPPING AND HANDLING
Prices vary depending on what you order and where you live.
See website or call for specifics.
To pay with check or money-order,
please call us first for address and shipping costs.

All products printed in full color on recycled paper with low VOC soy-based ink.

WORLD TIME ZONES

ID LW	NT BT	CA HT	YST	PST	MST	CST	EST	AST	BST	AT	WAT	GMT	CET	EET	BT	USSR Z3	USSR Z4	USSR Z5	SST	CCT	JST	GST	USSR Z10	ID LE
-12	-11	-10	-9	-8	-7	-6	-5	-4	-3	-2	-1	**0**	+1	+2	+3	+4	+5	+6	+7	+8	+9	+10	+11	+12
-4		-3	-2	-1	**0**	+1	+2	+3	+4	+5	+6	+7	+8	+9	+10	+11	+12	+13	+14	+15	+16	+17	+18	+19	+20

STANDARD TIME ZONES FROM WEST TO EAST CALCULATED FROM PST AS ZERO POINT:

IDLW:	International Date Line West	-4	**BT:**	Bagdhad Time	+11
NT/BT:	Nome Time/Bering Time	-3	**IT:**	Iran Time	+11 1/2
CA/HT:	Central Alaska & Hawaiian Time	-2	**USSR**	Zone 3	+12
YST:	Yukon Standard Time	-1	**USSR**	Zone 4	+13
PST:	Pacific Standard Time	0	**IST:**	Indian Standard Time	+13 1/2
MST:	Mountain Standard Time	+1	**USSR**	Zone 5	+14
CST:	Central Standard Time	+2	**NST:**	North Sumatra Time	+14 1/2
EST:	Eastern Standard Time	+3	**SST:**	South Sumatra Time & USSR Zone 6	+15
AST:	Atlantic Standard Time	+4	**JT:**	Java Time	+15 1/2
NFT:	Newfoundland Time	+4 1/2	**CCT:**	China Coast Time	+16
BST:	Brazil Standard Time	+5	**MT:**	Moluccas Time	+16 1/2
AT:	Azores Time	+6	**JST:**	Japanese Standard Time	+17
WAT:	West African Time	+7	**SAST:**	South Australian Standard Time	+17 1/2
GMT:	Greenwich Mean Time	+8	**GST:**	Guam Standard Time	+18
WET:	Western European Time (England)	+8	**USSR**	Zone 10	+19
CET:	Central European Time	+9	**IDLE:**	International Date Line East	+20
EET:	Eastern European Time	+10			

HOW TO CALCULATE TIME ZONE CORRECTIONS IN YOUR AREA:

ADD if you are **east** of PST (Pacific Standard Time); SUBTRACT if you are **west** of PST on this map (see right-hand column of chart above).

All times in this calendar are calculated from the West Coast of North America where We'Moon is made. Pacific Standard Time (PST Zone 8) is zero point for this calendar, except during Daylight Saving Time (March 10–November 3, 2019, during which times are given for PDT Zone 7). If your time zone does not use Daylight Saving Time, add one hour to the standard correction during this time. At the bottom of each page, EST/EDT (Eastern Standard or Daylight Time) and GMT (Greenwich Mean Time) times are also given. For all other time zones, calculate your time zone correction(s) from this map and write it on the inside cover for easy reference.

Conventional Holidays 2019

January 1	New Years Day*
January 21	Martin Luther King Jr. Day
February 5	Chinese/Lunar New Year
February 14	Valentines Day*
February 18	Presidents Day
March 6	Ash Wednesday
March 8	International Women's Day*
March 10	Daylight Saving Time Begins
March 12	Mexica New Year*
March 17	St. Patrick's Day*
April 14	Palm Sunday
April 19	Good Friday
April 20–April 27	Passover
April 21	Easter
April 22	Earth Day*
May 5	Cinco De Mayo*
May 6–June 4	Ramadan
May 12	Mother's Day
May 27	Memorial Day
June 16	Father's Day
July 4	Independence Day*
September 2	Labor Day
Sept. 30–Oct. 1	Rosh Hashanah
October 9	Yom Kippur
October 14	Indigenous Peoples' Day
October 31	Halloween*
November 1	All Saints' Day*
November 2	Day of the Dead*
November 3	Daylight Saving Time Ends
November 11	Veteran's Day*
November 28	Thanksgiving Day
Dec. 23–30	Chanukah/Hanukkah
December 25	Christmas Day*
December 26	Boxing Day*
Dec. 26–Jan. 1	Kwanzaa*
December 31	New Years Eve*
*Same date every year	

Become a We'Moon Contributor!

Accepting submissions for
We'Moon 2021: the 40th edition!
Tentative Theme: The World

Call for Contributions: Available in the spring of 2019

Postmark-by Date for all art and writing: August 15, 2019

Note: It is too late to contribute to

We'Moon 2020: Wake Up Call

We'Moon is made up by writers and artists like you! We welcome creative work by women from around the world, and aim to amplify diverse perspectives. We especially encourage those of us who are women of color or who are marginalized by the mainstream, to participate in helping We'Moon reflect our unique visions and experiences. We are eager to publish more words and images depicting people of color created by WOC. By nurturing space for all women to share their gifts, we unleash insight and wisdom upon the world—a blessing to us all.

> **We invite you to send in your art and writing for the next edition of We'Moon!**

Here's how:

Step 1: Visit wemoon.ws to download a Call for Contributions or send your request for one with a SASE (legal size) to **We'Moon Submissions, PO Box 187, Wolf Creek, OR 97497.** (If you are not within the US, you do not need to include postage.) The Call contains current information about the theme (it may change), specifications about how to submit your art and writing, and terms of compensation. There are no jury fees. The Call comes out in the early Spring every year.

Step 2: Fill in the accompanying Contributor's License, giving all the requested information, and return it with your art/writing and a self addressed envelope by the due date. *No work will be accepted without a signed license!*

Step 3: Plan ahead! To assure your work is considered for *We'Moon 2021*, get your submissions postmarked by August 15, 2019.

NOTES

Jewel of the Night
© *Tamara Phillips 2017*

Lingering Trees
© *Janis Dyck 2014*

Sun Salutation
© *Jenny Hahn 2004*

The Huntresses. Reflection.
© *Paula Franco 2017*

239

Ephemeral Focus
© Autumn Skye ART 2015

A Tea Offering
© Sarah Cook 2016